室内细部图集 3
办公场所与教育机构

INTERIOR DETAIL, OFFICE & EDUCATIONAL INSTITUTION

凤凰空间 编

细部节点 100例

江西科学技术出版社

Contents

餐厅

商店与
住宅

办公场所与
教育机构

咖啡馆

医疗中心与
文化中心

酒店与
休闲场所

*本书中所有无单位数值均为毫米。

办公场所

现代Hmall	6
VetAll公司	20
GWANGJU创意经济革新中心	30
韩华梦想+BI中心	40
哈拿多乐旅行社高级路边店	56
Daum Games办公场所	66
阿法拉伐亚洲综合办事处	76
SMART MEDIA创新中心	86
Kkotsbom办公场所	94
Riot Games办公场所	106
Tapjoy办公场所	122

教育机构

DongEun-I幼儿园	130
Presby教育中心	144
KUT 环球教育中心	162
人寿保险	172

办公场所

- 现代Hmall
- VetAll公司
- GWANGJU创意经济革新中心
- 韩华梦想+BI中心
- 哈拿多乐旅行社高级路边店
- Daum Games办公场所
- 阿法拉伐亚洲综合办事处
- SMART MEDIA 创新中心
- Kkotsbom办公场所
- Riot Games办公场所
- Tapjoy办公场所

| 现代Hmall | 建筑面积：26 001平方米

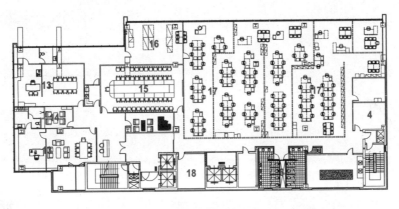

1. 咖啡间
2. 洗漱间
3. 自助餐厅
4. 仓库
5. 厨房
6. 厕所
7. 信息台
8. 大厅
9. 舞台
10. 露天休息区
11. 洽谈室
12. 休息区
13. 行政套房
14. 接待室
15. 会议室
16. 实验室
17. 办公室
18. 咨询室

九层平面图

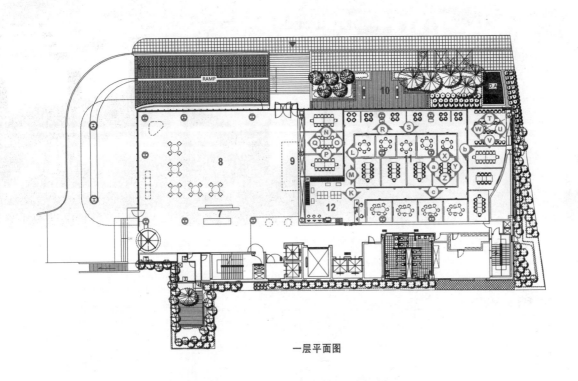

一层平面图

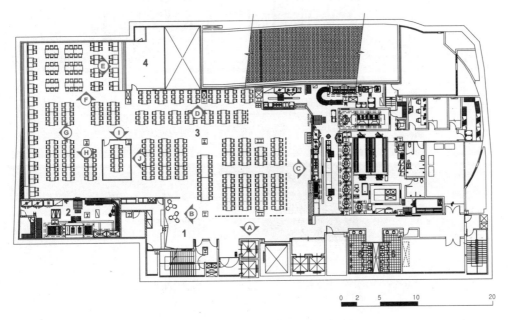

地下一层平面图

现代Hmall

■ 地下一层餐厅

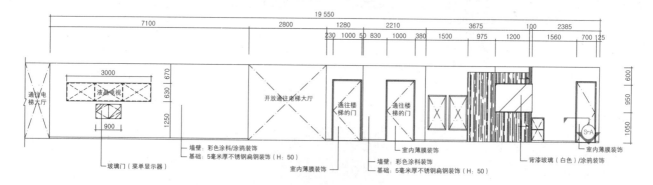

立面图A

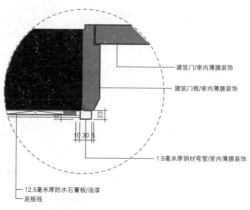

剖面图A

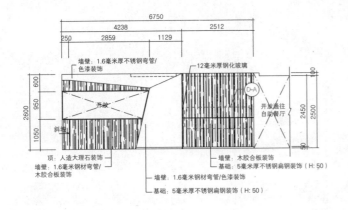

立面图B

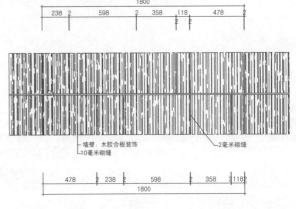

细节图A

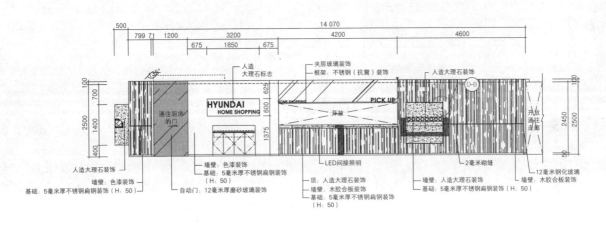

立面图C

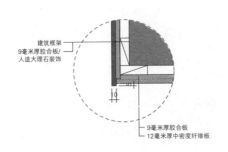

细节图B

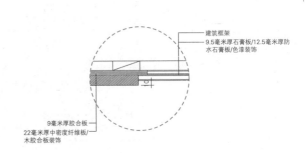

剖面图B

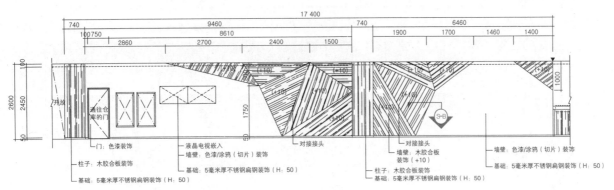

立面图D

现代Hmall

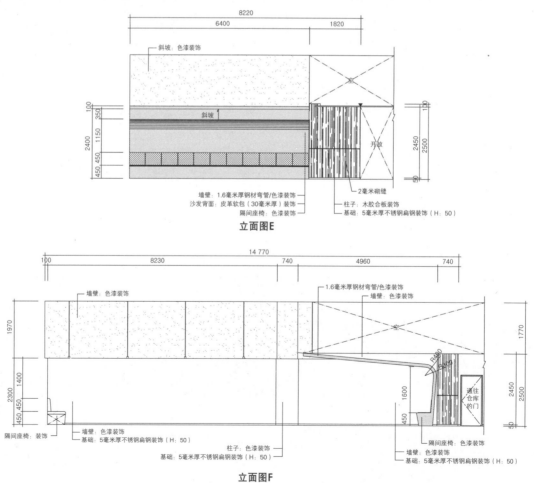

立面图E

立面图F

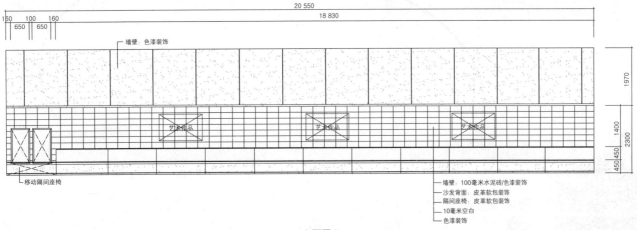

立面图G

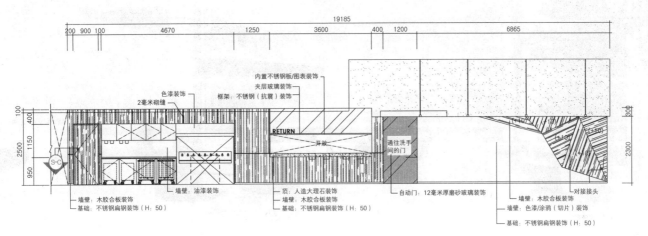

立面图H

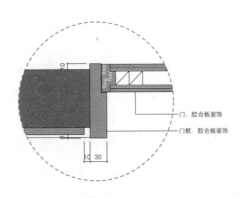

剖面图C

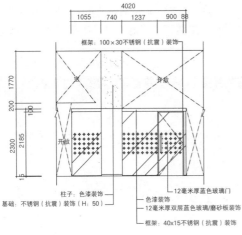

立面图I 　　　　　立面图J

The office has gone under a change into a high-tech environment in response to the company's continuous growth through remodelling work, which involves vertical and horizontal expansion using unprecedented slanted planes. Two new floors were added to the original thirteen floor building; one of them is rooftop garden featuring a green zone for relaxation. The company canteen on the 10th floor was relocated to the 1st floor and completely refurbished into a high quality restaurant. Considering that the broadcasting's sensitivity to noise and vibration, the designer enhanced the earthquake-proof performance and applied advanced construction methods while creating a cozy and pleasant atmosphere for the workspace, meeting rooms, and convenient facilities. The performance was also improved by the introduction of water recycling system, which uses gray water for sundry purposes, and solar energy generating device, thus increasing the sustainability of the building.

Through complete renovation, the office would elevate the pride of the staff in the corporate.

为了应对公司的持续增长，办公室经过重新设计改造已经变成一处高科技场所，该设计采用前所未有的倾斜板实现纵向和横向扩建。建筑物在原来十三层的基础上另扩建两层；其中一层为屋顶花园，作为休闲的绿色地带。第十层的公司餐厅搬迁至第一层，被彻底翻新成一家高档饭店。考虑到广播对噪声和震动的敏感性，设计师采用了先进的施工方法，提高了降噪和抗震性能，同时为工作场所、小会议室和便利设施营造舒适愉悦的氛围。另外引进了水循环系统，将中水用于日常生活和太阳能发电设备，进一步提高建筑物的可持续性。

通过彻底的翻修，办公室将提高公司员工的自豪感。

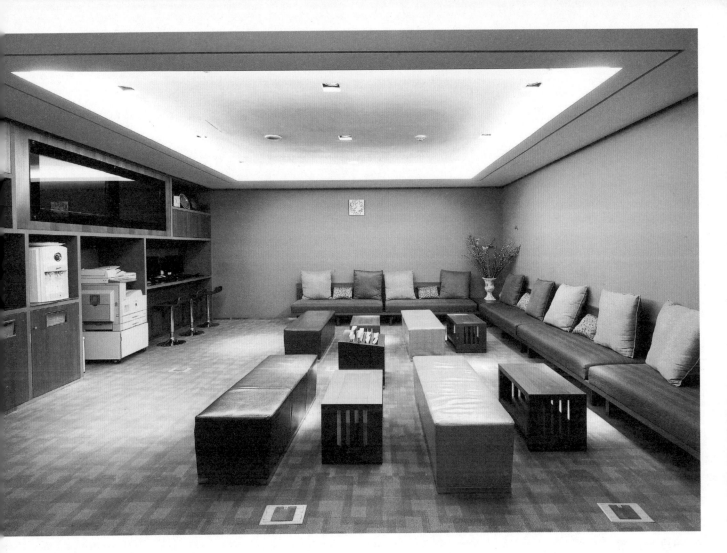

一层休息区

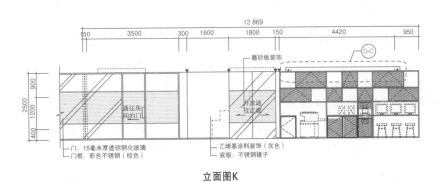

立面图K

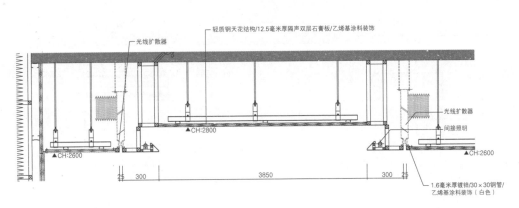

细节图C

■ 走廊

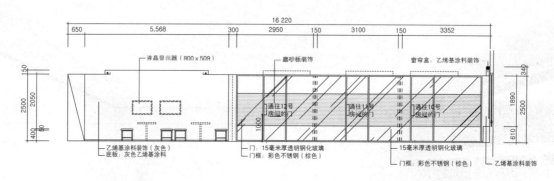

立面图L

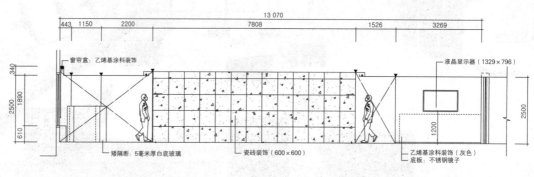

立面图M

■ 一层洽谈室

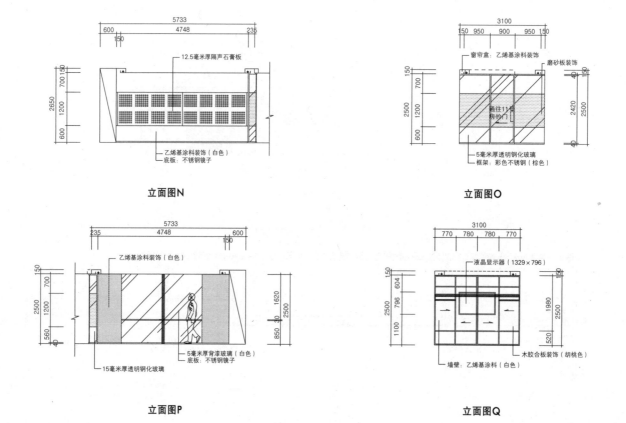

立面图N　　　　　　　　　立面图O

立面图P　　　　　　　　　立面图Q

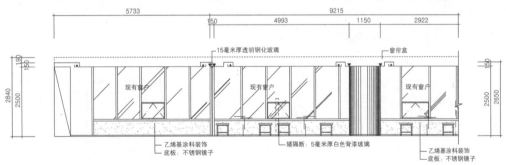

立面图R

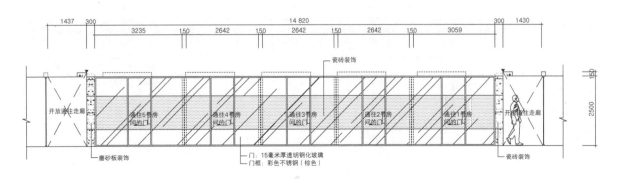

立面图S

现代Hmall

■ 洽谈室

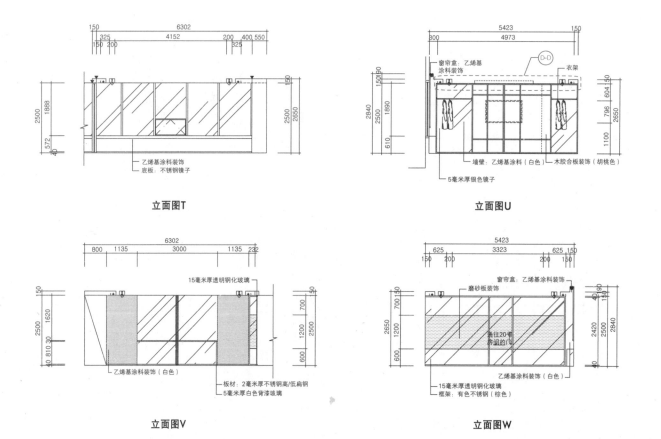

立面图T

立面图U

立面图V

立面图W

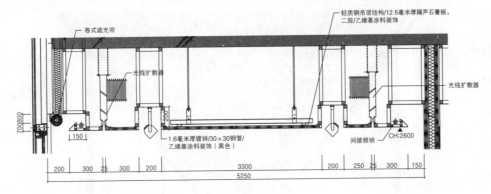

细节图D

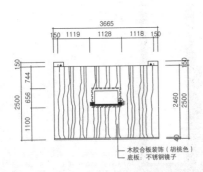

立面图X

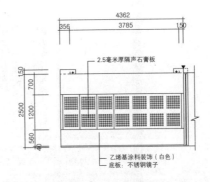

立面图Y

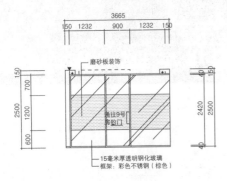

立面图Z

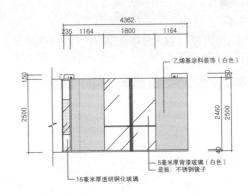

立面图a

■ 会议桌

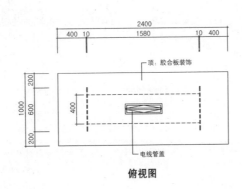

俯视图

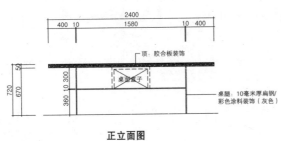

正立面图

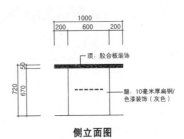

侧立面图

■ 洽谈区走廊

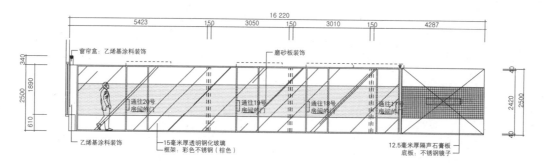

立面图b

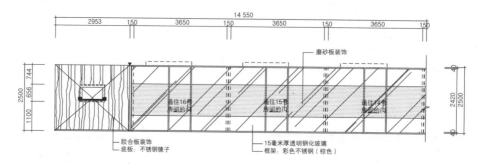

立面图c

正面外观

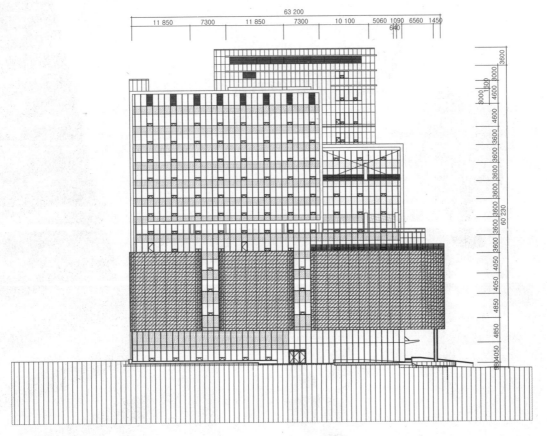

现代Hmall 19

VetAll公司

建筑面积：（一层）31平方米，（二层）484平方米

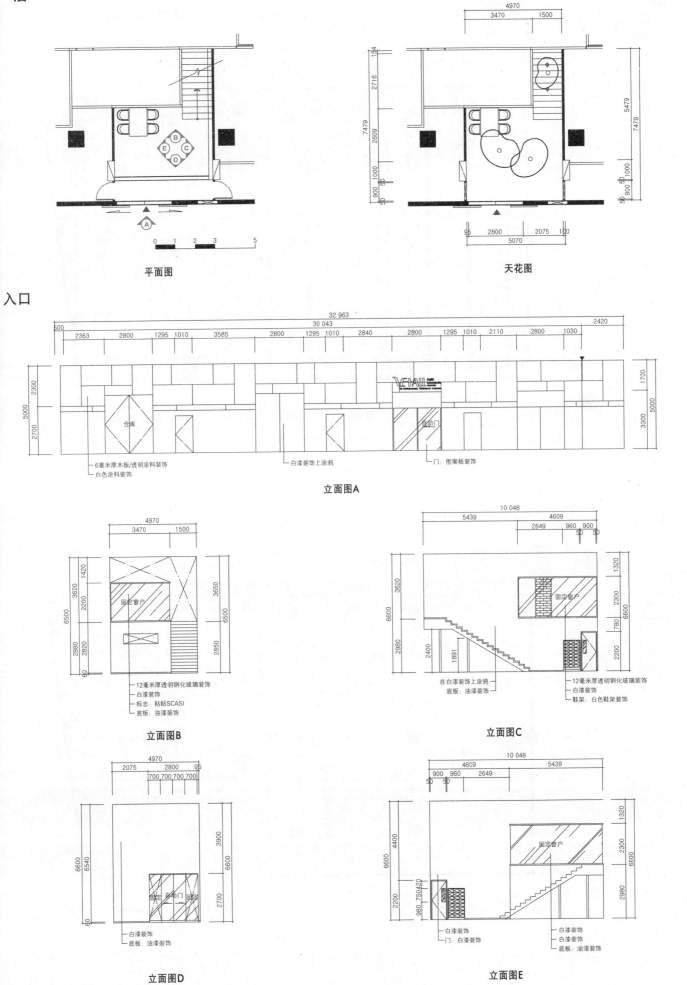

二层

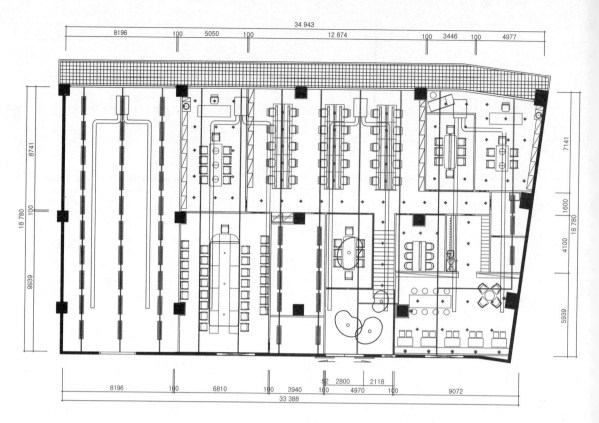

二层天花图

1. 入口
2. 走廊
3. 办公室
4. 总裁办公室
5. 会议室
6. 仓库
7. 接待室
8. 休息区
9. 图书室
10. 执行董事办公室

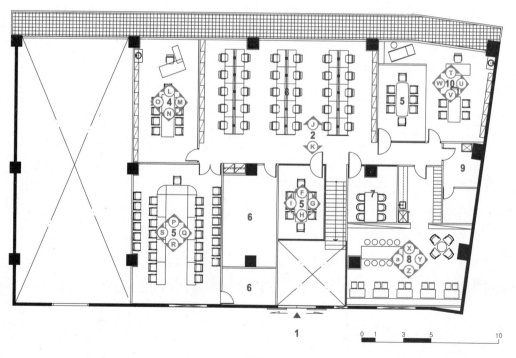

二层平面图

VetAll Laboratories, which develops and manufactures diagnostic products for animal health, has been designed as a clean and comfortable space that breaks away from the typical rigid atmosphere. The colors of the sky and nature fill the spacious area.

The building consists of two floors; the first floor is occupied by a lobby, and the second floor by a work space and a meeting room. The two-level-high lobby is characterized by the lighting shaped as a cloud, and by the green color of the logo that delivers a sense of unity to the overall interior. The designer exposed the ceiling of the second floor to compensate the rather small space crowded with office furniture, and applied green color to ventilator openings to tidy up the entangled equipments. The meeting rooms in places were separated by whole glass to make the interior look more spacious, which is accentuated by the OSB veneer and the colors of sky blue and light green. The break area presents an informal and casual atmosphere of a cafe. Contrasting to the white office area, it radiates a vintage look by the use of old bricks and the huge pendant, thus offering the users a cozy and comfortable relaxation area filled with soothing energy.

研发并生产动物保健产品的 VetAll 实验室，被设计成清洁舒适的空间，打破了传统死板的气氛。宽敞的空间里用天空和自然的颜色装饰。

该建筑物包括两层：第一层为大厅，第二层为工作区和会议室。两层高的大厅采用云朵状的灯具，绿色的标志使整个室内空间的风格得以统一。设计师将第二层天花板裸露在外，弥补了办公室空间小且挤满办公家具的缺陷；绿色的通风孔使相关设备显得比较整洁。会议室用全玻璃隔开，使空间看起来更宽敞，OSB 胶合板、天蓝色和浅绿色的选用更凸显这一点。隔断区呈现出咖啡馆式的轻松休闲氛围，与白色办公区不同，它采用旧式砖和硕大的吊灯，流露出一种复古的风格，为使用者打造了一处轻松舒适且充满舒缓气息的休闲空间。

■ 会议室

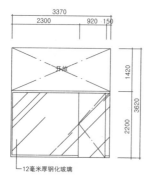

立面图F

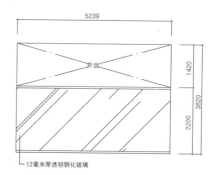

立面图G

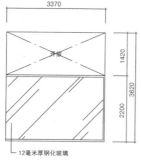

立面图H

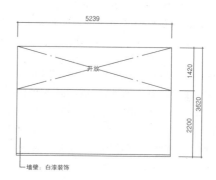

立面图I

办公室

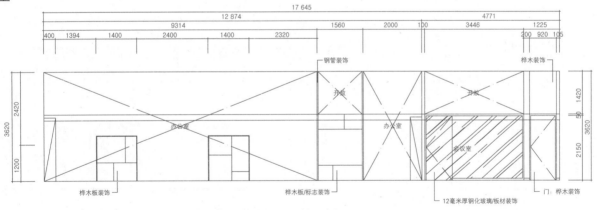

立面图 J

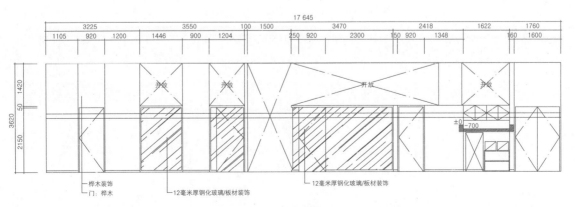

立面图 K

总裁办公室

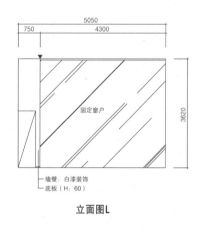

立面图L

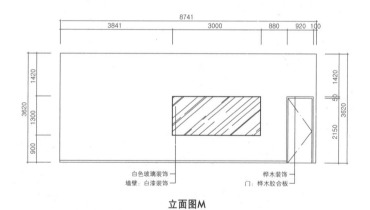

立面图M

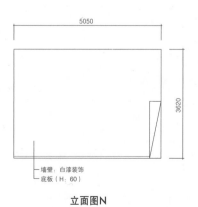

立面图N

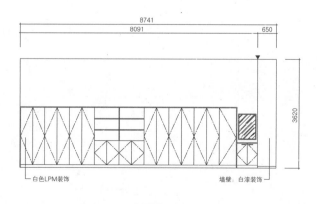

立面图O

大会议室

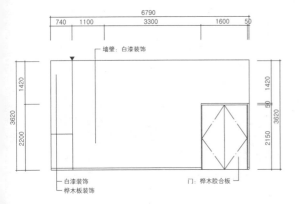

立面图P

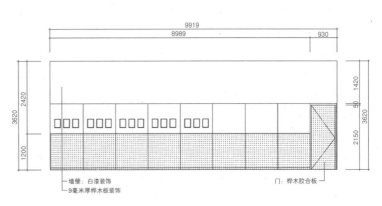

立面图Q

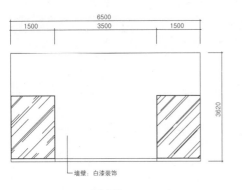

立面图R

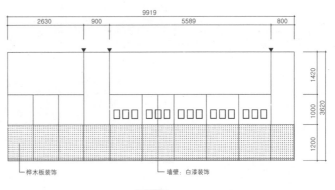

立面图S

■ 执行董事办公室

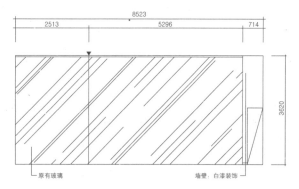

立面图T

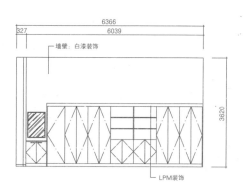

立面图U

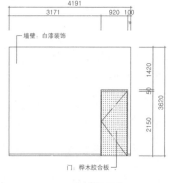

立面图V

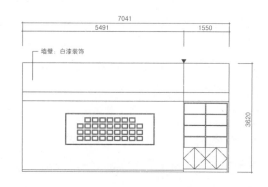

立面图W

■ 休息区

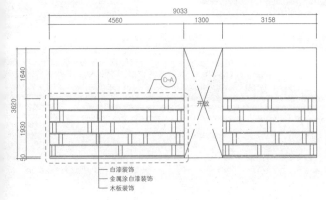

立面图X

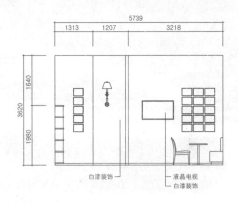

立面图Y

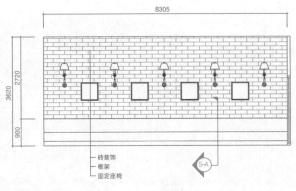

立面图Z

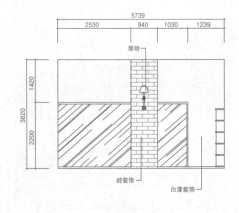

立面图α

■ 细节图A——书架

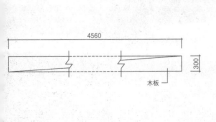

俯视图

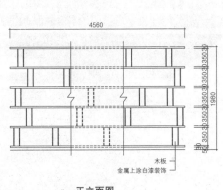

正立面图

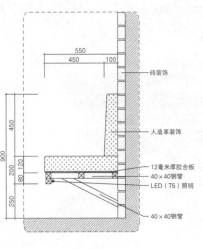

剖面图A

|GWANGJU创意经济革新中心|

建筑面积：1214平方米

外部

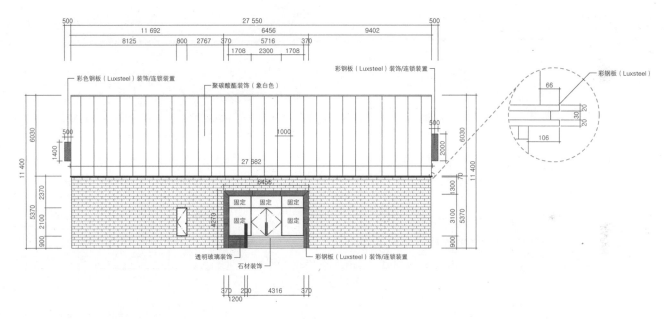

正立面图

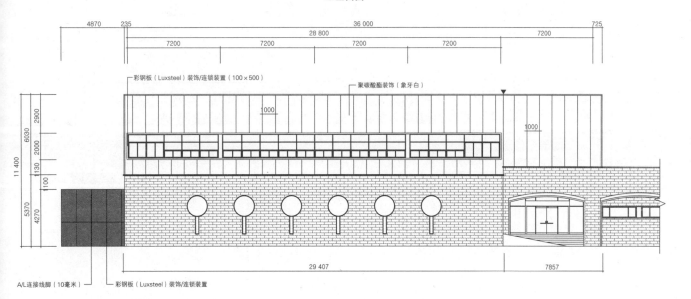

侧立面图

GWANGJU创意经济革新中心

1. 入口
2. 咨询台
3. 开放式会议区
4. 会议室
5. 办公室
6. 服务中心
7. 样板间
8. 展示区
9. 大厅
10. 团队工作室

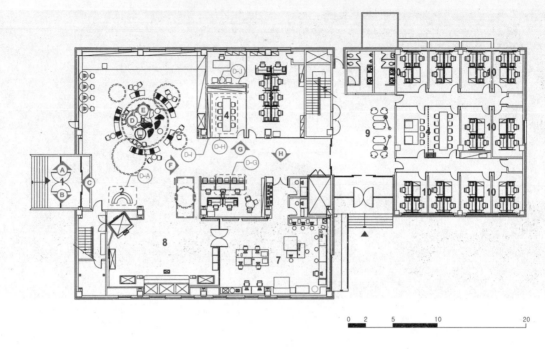

平面图

■ 入口

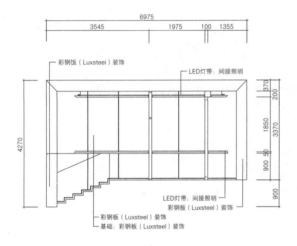

立面图A

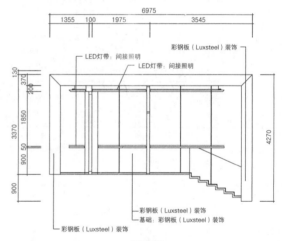

立面图B

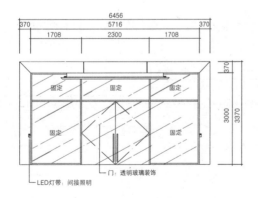

立面图C

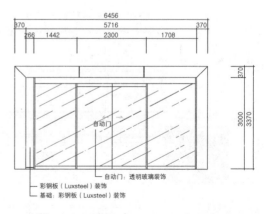

立面图C'

细节图A——咨询台

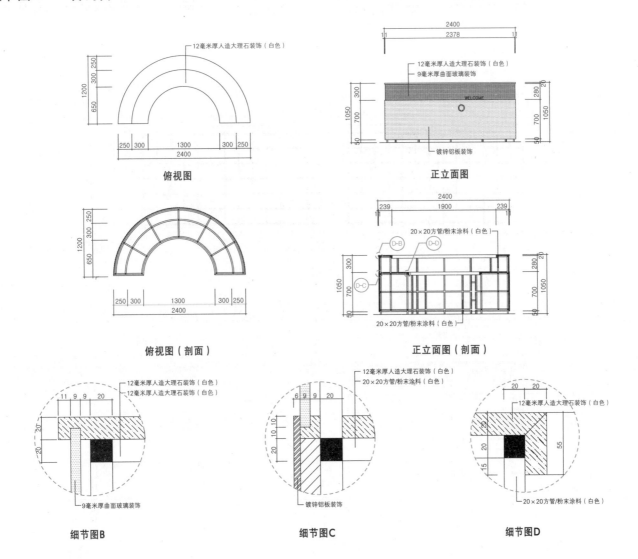

俯视图

正立面图

俯视图(剖面)

正立面图(剖面)

细节图B

细节图C

细节图D

GWANGJU创意经济革新中心

开放式会议区

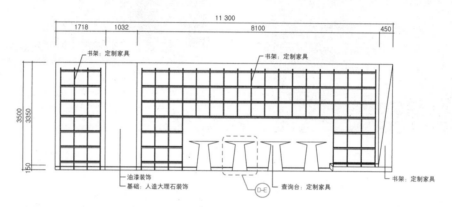

立面图D

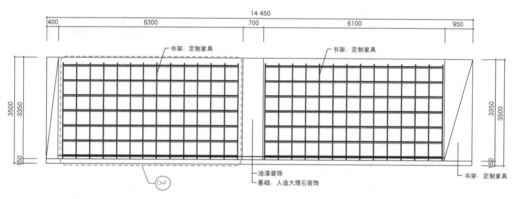

立面图E

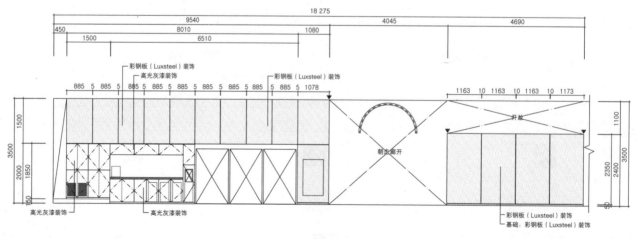

立面图F

细节图E——查询台

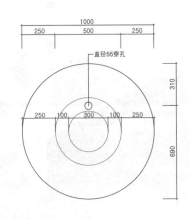

俯视图

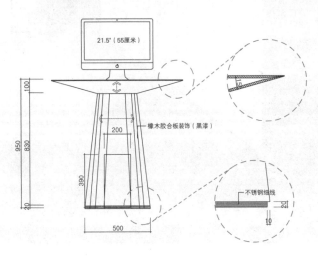

正立面图

细节图F——书架

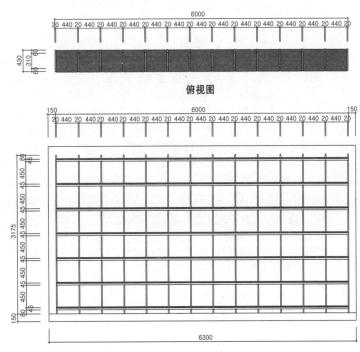

俯视图

正立面图

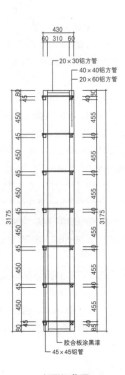

剖面细节图

GWANGJU创意经济革新中心　35

The underused space in Gwangju Institute of Science and Technology was renovated into a business establishment and research facility related to hydrogen energy and hydrogen fueled vehicle.

As the center of the institute, the conference zone was designed as a communicative area open not just to students but also to general public. The designer minimized the walls and applied movable furniture to ensure the space for various meetings and lectures. The arcade beyond the hall is different in height and finish materials, thus separating itself from others and at the same time guide the visitors to the inside. At the furthest back are the research room and conference room for venture companies, where the users can focus on research and development in a cozy and quiet atmosphere, away from the noise from the outside.

The selected materials such as metal, plastic, carpet, polycarbonate, and concrete block do not overwhelm the interior and exterior but rather emphasize their quality. The hexagonal patterns on the flooring, walls, and other elements remind the hydrogen molecule, thus reinforcing the identity of the space.

GWANGJU 科学技术学院未利用的空间被整修成一个与氢能和氢燃料汽车相关的商业机构和研究场所。

作为该学院的创意经济革新中心，会议区被设计成一个交流区域，对学生和公众开放。设计师尽量减少墙壁，并采用可移动家具，确保各种会议和演讲具有足够的空间。大厅外的拱廊高度和装饰材料与其他区域不同，从而区分开来，同时引导参观者入内。最后面是创业公司的研究室和会议室，这里使研究人员可以在舒适安静的氛围中专心研发，远离外界噪声等干扰。

金属、塑料、聚碳酸酯和混凝土砖等材料并不会影响室内外的布景，反而凸显了它的品质。地板、墙壁和其他组成部分上的六边形图案令人联想起氢分子，从而强化了空间的个性。

走廊

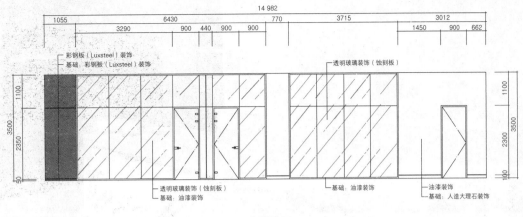

立面图G

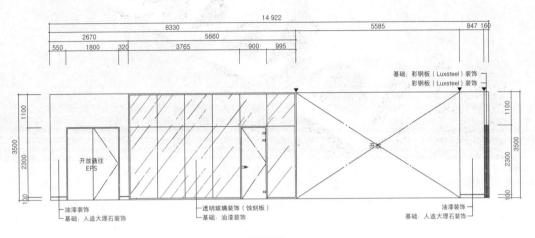

立面图H

细节图G——服务台

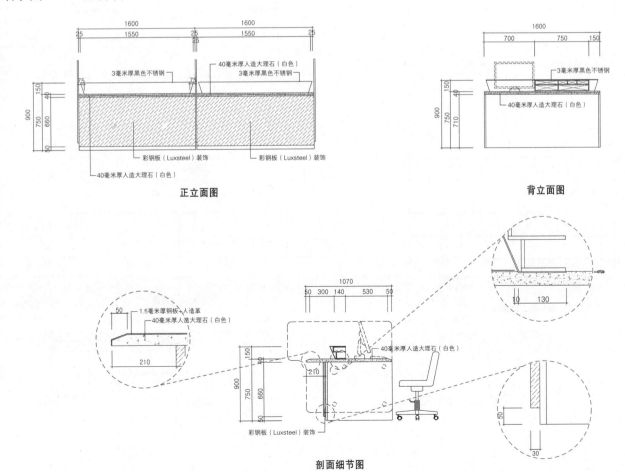

正立面图　　　　　　　　　背立面图

剖面细节图

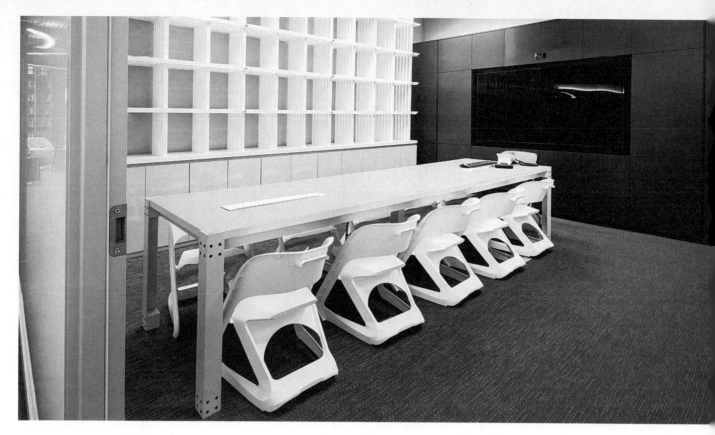

细节图H——会议室的桌子

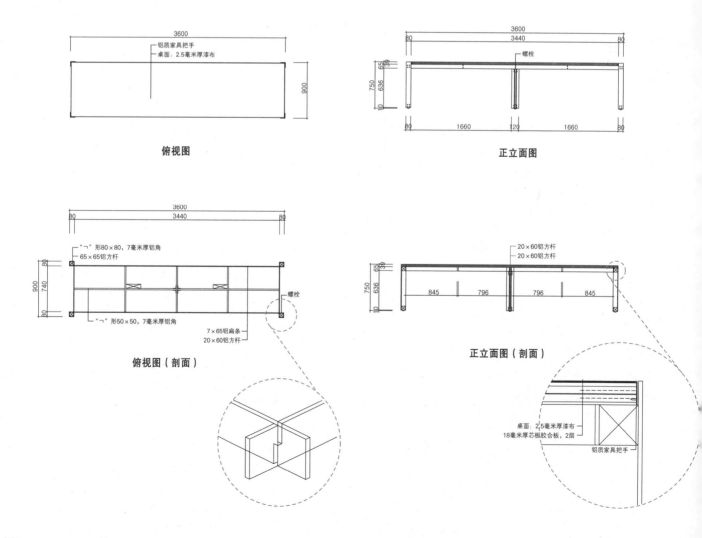

细节图I——书架

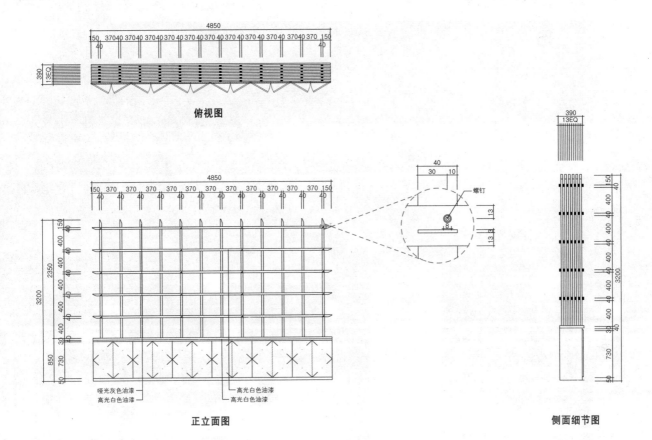

细节图J——固定家具

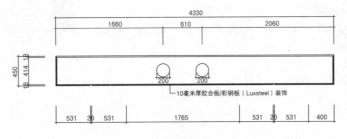

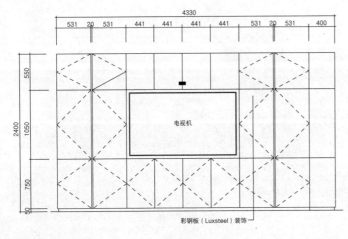

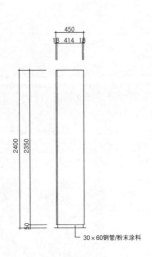

韩华梦想+BI中心

建筑面积：1540平方米

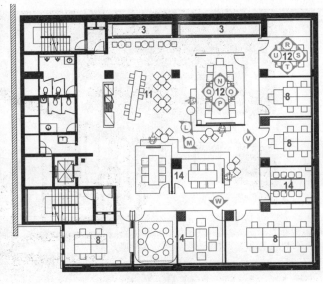

地下一层平面图

1. 大厅
2. 厕所
3. 下沉区
4. 图书馆
5. 仓库
6. 攀岩区
7. 瑜伽区
8. 办公区
9. 乒乓球区
10. 影印室&餐厅
11. 休息区
12. 创意团队工作室
13. 浴室
14. 会议室
15. 经理会议室
16. 经理办公室
17. 经理休息室
18. 平台

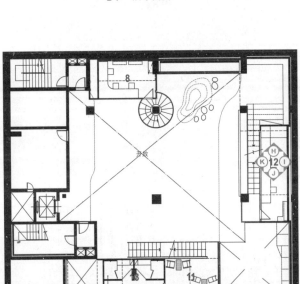

地下二层平面图

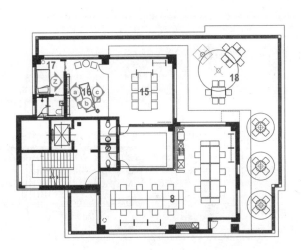

四层平面图

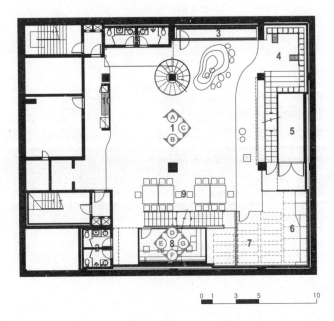

地下三层平面图

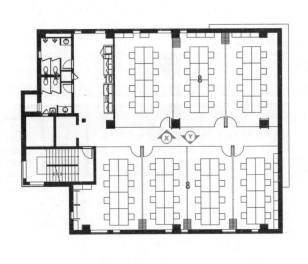

二层平面图

地下三层大厅

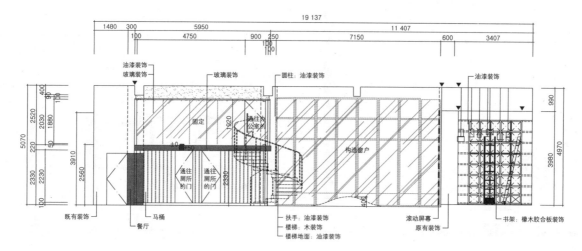

立面图A

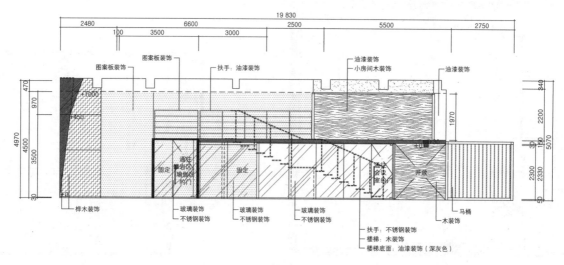

立面图B

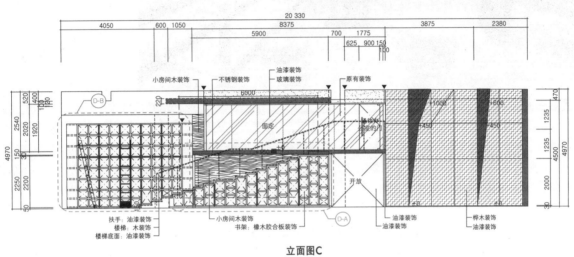

立面图C

■ 细节图A

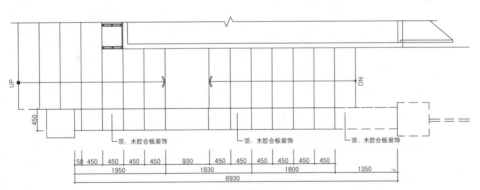

俯视图

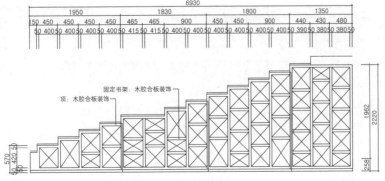

正立面图

韩华梦想+BI中心

■ 细节图B——书架

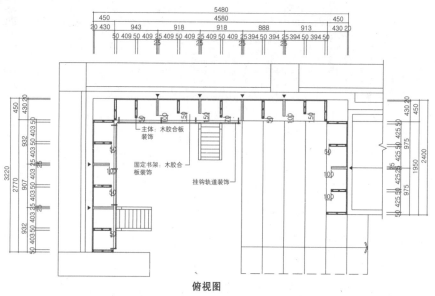

俯视图

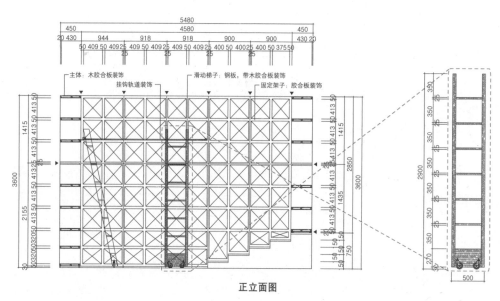

正立面图

44

地下三层办公室

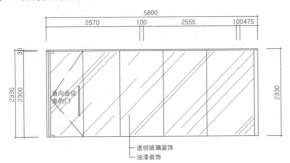

立面图D

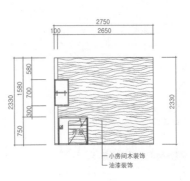

立面图E

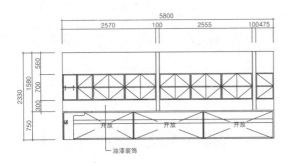

立面图F

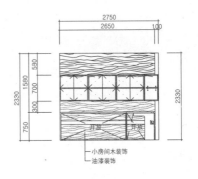

立面图G

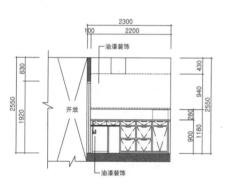

立面图H

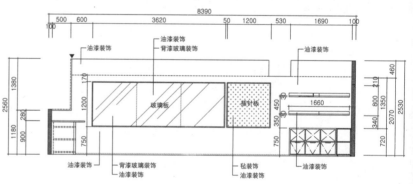

立面图I

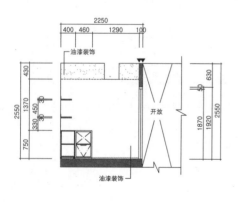

立面图J

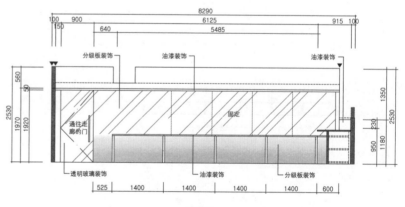

立面图K

Hanwha Dream Plus BI Center has opened to support 'Start Up' program that fosters small companies.

Consisting of three underground floors and four above-ground floors, the center is themed with 'Dream Plant', interpreting the flow of time into lines in a creative way. The third underground floor houses functional and dynamic areas such as a hall, library, climbing facility, ping pong table, pantry, and a shower room. The natural elements are strewn over the place under the two-level ceiling, looking like a forest. The first underground floor is occupied by various meeting spaces with the concept of a 'square'; the themes of Korea, Rome, Mongol, Caribbean, and Space create vitality that encompasses time and space. In the offices from the second to fourth floor, the partitions on the wall allow the flexibility in the space. The courtyard and a terrace on the fourth floor provide an open and liberal work space.

Hanwha Dream Plus BI Center offers diverse companies to work in the creative and characteristic space.

韩华梦想+BI中心是一家专注扶植小型公司起步创业项目的公司。

该办公楼由地下三层和地上四层构成，主题为"梦想工厂"，以创新的方式将时间转变成生产线。地下第三层设有大厅、图书室、攀岩设施、乒乓球台、配餐室和浴室等功能区和动态区。第二层的天花板下布满自然元素，看上去像一片森林。地下一层布置以"广场"为理念打造的各种会议空间；以韩国、罗马、蒙古、加勒比和太空为主题营造出时空的活力。第二至第四层为办公室，墙壁上的隔断实现了空间的灵活性。第四层上的庭院和阳台提供了一处开放、自由的工作区。

韩华梦想+BI中心为各种公司提供了富有创意和特色的工作空间。

■ 地下一层走廊

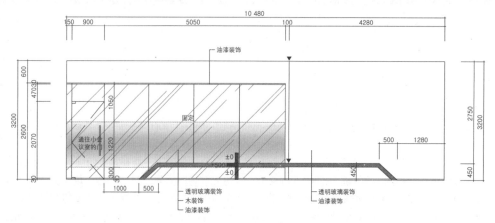

立面图L

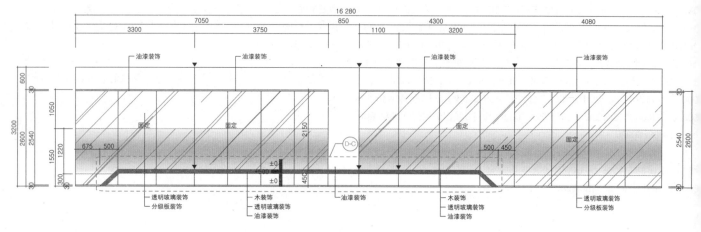

立面图M

细节图C——长凳

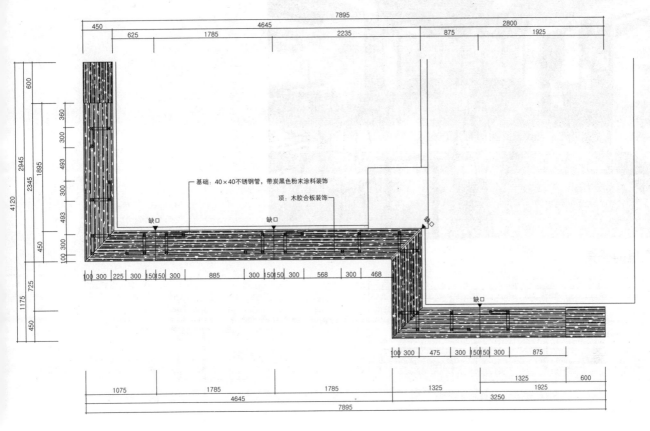

俯视图

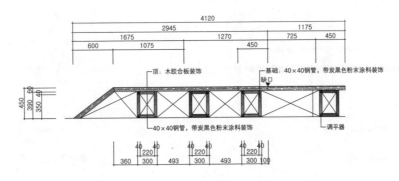

正立面图

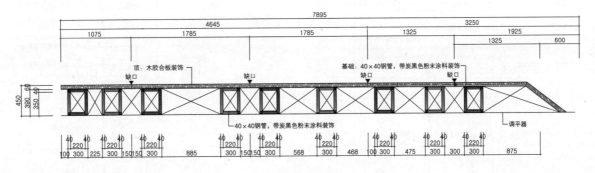

正立面图

休息区里的长桌

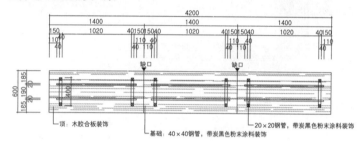

俯视图

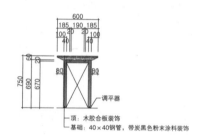

侧立面图

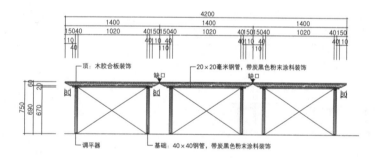

正立面图

侧立面图（剖面）

休息区餐厅

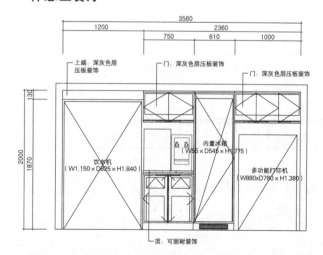

正立面图

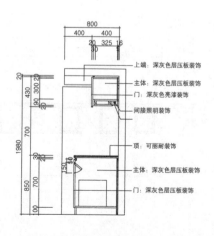

侧立面图（剖面）

岛屿会议室

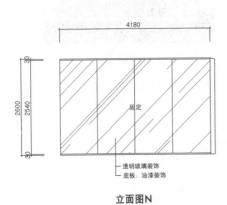

立面图N

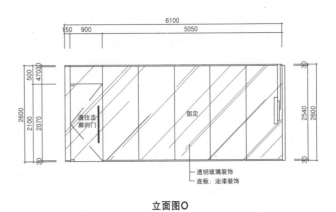

立面图O

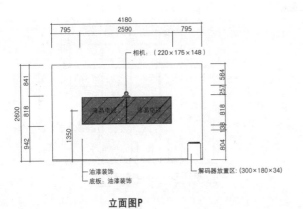

立面图P

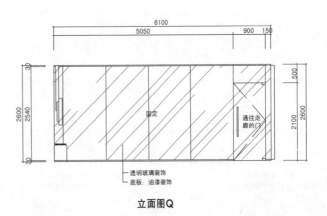

立面图Q

会议室——罗马主题

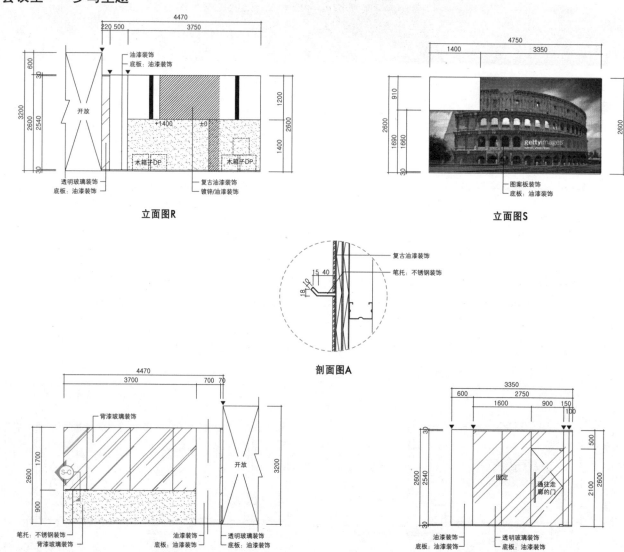

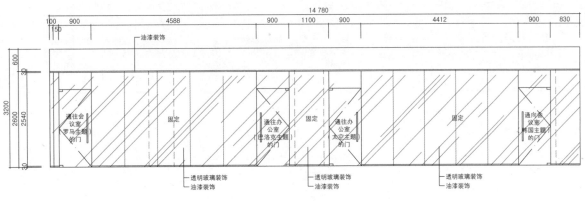

立面图V

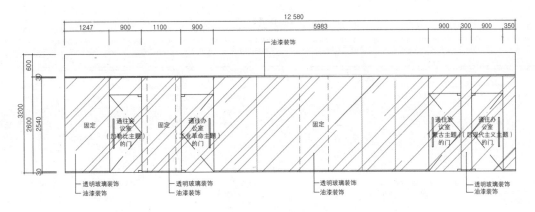

立面图W

■ 二楼走廊

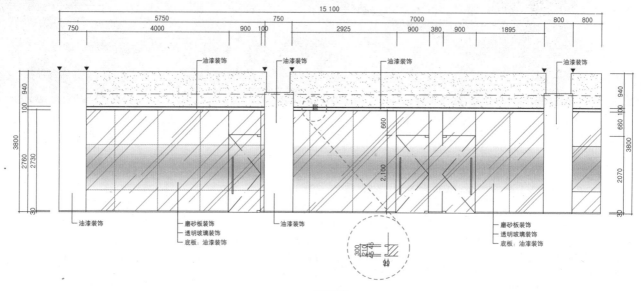

立面图X

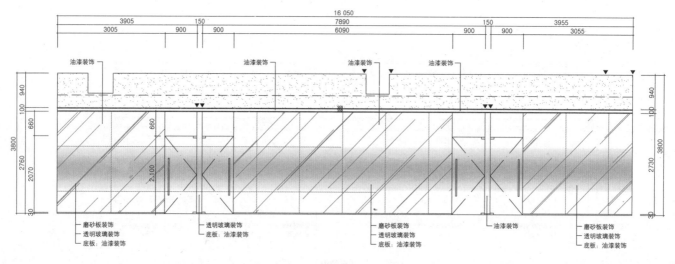

立面图Y

经理办公室

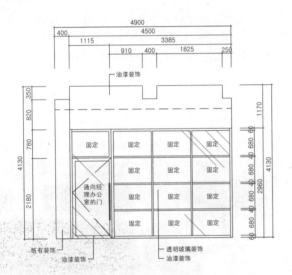

立面图Z

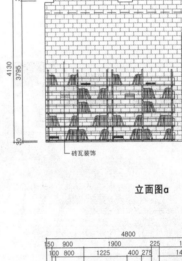

立面图a

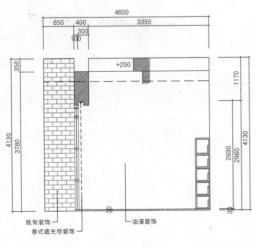

立面图b

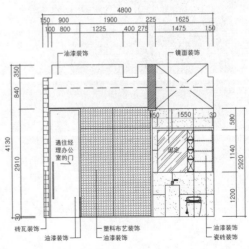

立面图c

韩华梦想+BI中心 55

哈拿多乐旅行社高级路边店

建筑面积：314平方米

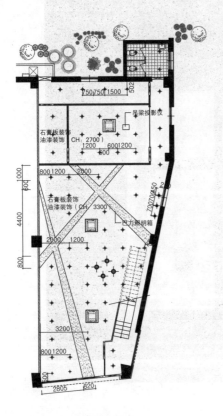

第一套天花图

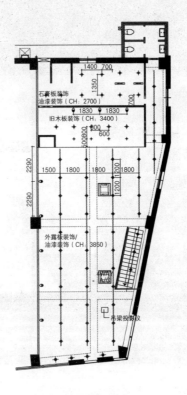

第二套天花图

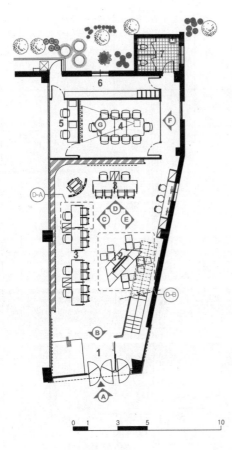

第一套平面图

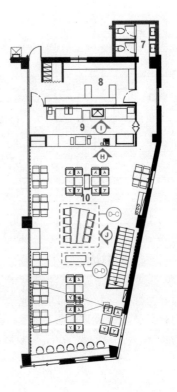

第二套平面图

1. 入口
2. 休息区
3. 咨询台
4. 会议室
5. 办公室
6. 仓库
7. 厕所
8. 员工更衣橱
9. 厨房
10. 咖啡馆
11. 研讨区

■ 正面外观

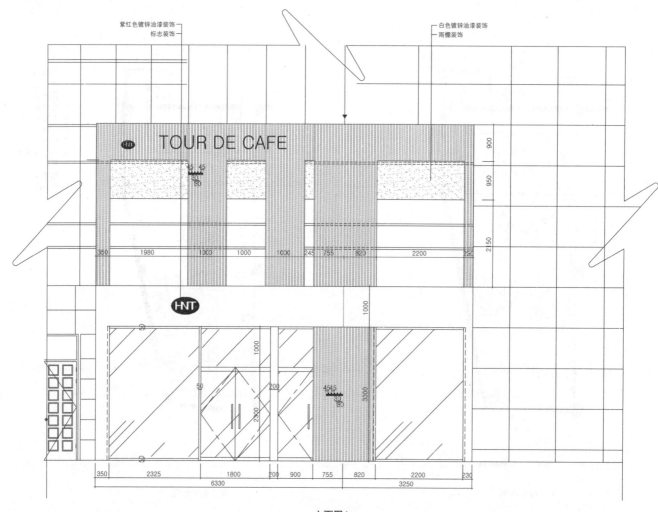

立面图A

■ 入口

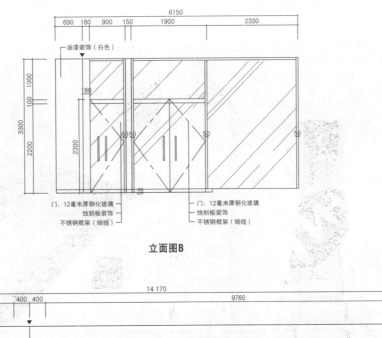

立面图B

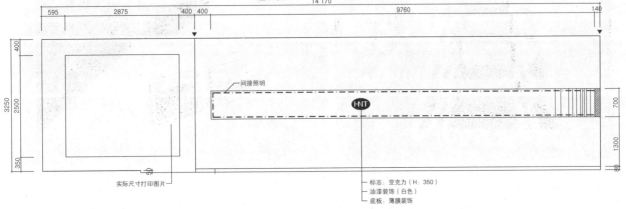

立面图C

一层旅行社

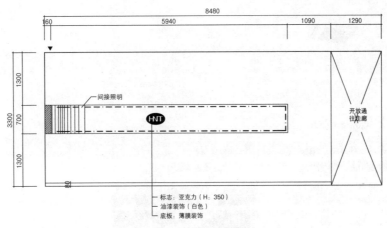

立面图D

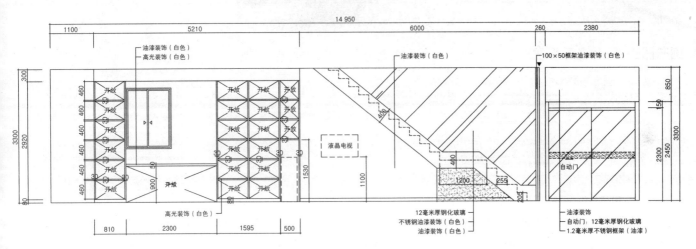

立面图E

细节图A——咨询台

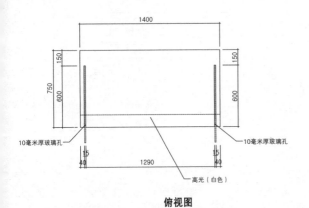

俯视图

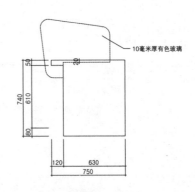

侧立面图

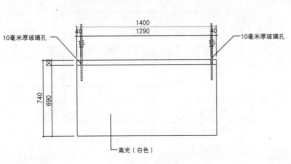

正立面图

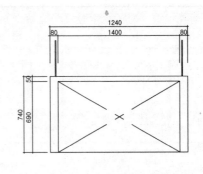

背立面图

■ 细节图B——休息区沙发

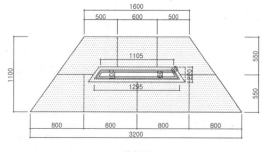

俯视图

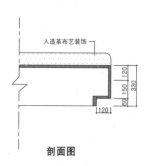

剖面图

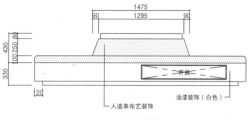

正立面图

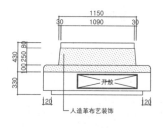

侧立面图

■ 会议室

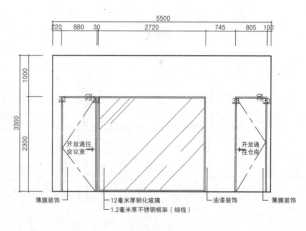

立面图F

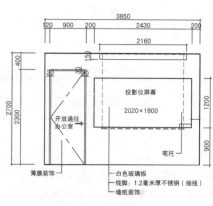

立面图G

二楼咖啡馆

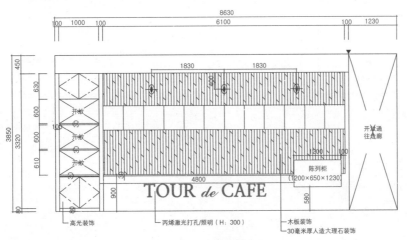

立面图H

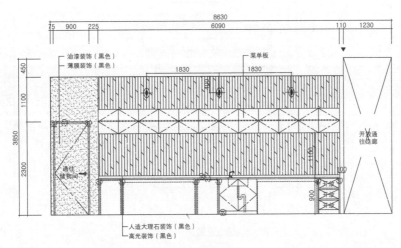

立面图I

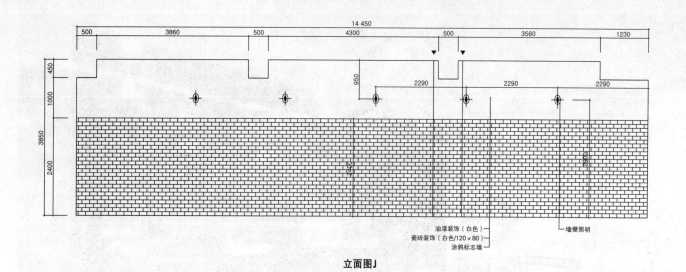

立面图J

细节图C——咖啡桌

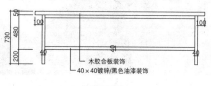

正立面图

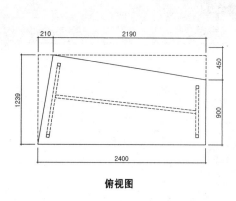

俯视图

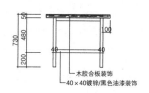

侧立面图

细节图D——书架

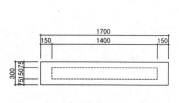

俯视图

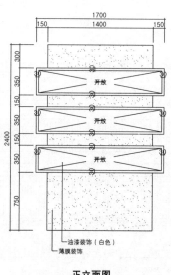

正立面图

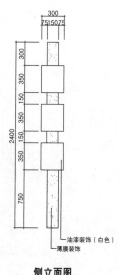

侧立面图

At the entrance is a large framed photograph that stimulates the anticipation of a trip. On the inside, the diagonal pattern on the ceiling offsets the possible monotony resulting from the series of consultation booths, and gives vitality to the space. The flooring, walls, ceiling, and furniture are all finished in white in order to emphasize that the space is dedicated to the newlyweds. The staircase decorated with various trip-related graphics leads to a cafe. The typography, design furniture, and accessories in the second floor cafe represent the most visited cities in the world by people in their twenties and thirties, catching the atmosphere of tourist destinations under the concept of 'travel'. The table on the center serves various purposes such as providing tourist information, giving presentations on a beam project as well as emphasizing the function as a complex cultural space.

Blending the boundary between a travel agent and a cafe, Hana Tour offers a public space for those who plan/want to go on a trip, or have an interest in trips abroad.

入口处有一大幅照片，激起人们对旅行的期待。进入室内，天花板上的对角图案弥补了一系列咨询台可能产生的单调感，赋予空间活力。地板、墙壁、天花板和家具均采用白色，以强调该场所是专为新婚人士打造的。通向咖啡厅的楼梯用各种旅行图案装饰。二楼咖啡厅的印刷、设计家具和装饰展示了二三十岁的年轻人最常旅行的世界各国的城市，在"旅行"的概念下再现旅游目的地的氛围。中间的桌子具有多种用途，比如提供旅行信息并在投影仪上做演示，更像是一个多功能的文化场地。

这间店模糊了旅行社与咖啡厅的界限，为计划或希望旅行，或者有兴趣前往国外旅行的人提供了一处特别场所。

Daum Games办公场所

建筑面积：1357平方米

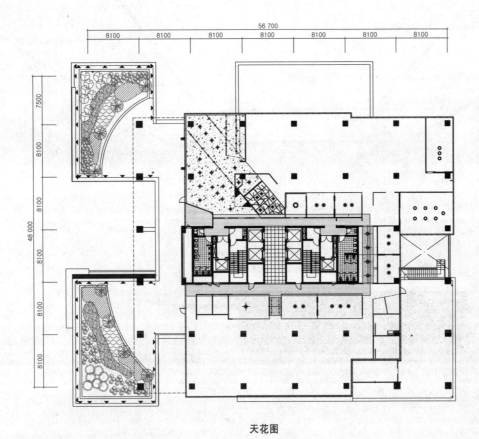

天花图

平面图

1. 电梯厅
2. 咨询台
3. 游戏室
4. 自助餐厅
5. 接待室
6. 会议室
7. 办公室
8. 总裁办公室
9. 大会议室
10. 餐厅
11. 储物柜
12. 卧室
13. 仓库
14. 问答室

Daum Games办公场所　67

Under the concept of 'chic Black & casual Wood', the project is defined by its unique selection of color and material. The slanted front desk and wire netting on the ceiling, fluorescent lighting installed in a free form present the chic and resolute identity of the company. The cafeteria finished in wood creates contrasting warmth in the back. the small conference room and break zone are vitalized by fabric furniture and pendant lightings of various designs and pastel colors, while the grand conference room and corridor to the locker room are covered in achromatic colors to give a still and solemn ambience.

The unique office space shows the harmony of the liberal mind and resolute spirit of the staff.

在"别致黑和随意木"的概念下，该项目因其独特的颜色和材料而独树一帜。倾斜的前台、天花板上的铁丝网以及自由布置的荧光灯展现出公司别致、果敢的特征。位于后面的自助餐厅使用木材来装饰，显得更加温暖。小型会议室和休息区采用各种造型及粉色系布艺家具和吊灯，富有生气，而大会议室和通向衣帽间的走廊采用单色，形成静谧庄重的氛围。

该独特的办公场所凸显了员工自由的思想和果敢的精神。

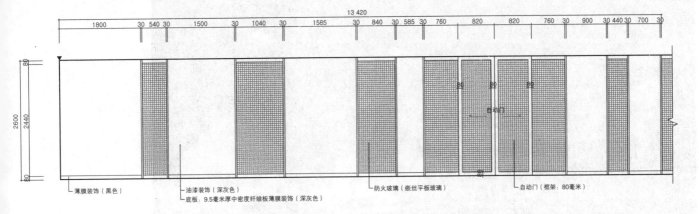

立面图A

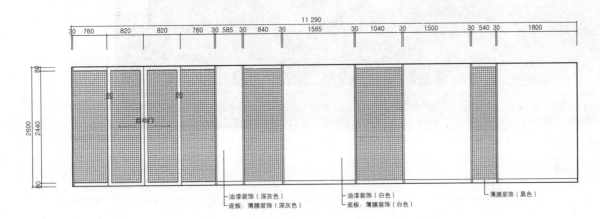

立面图B

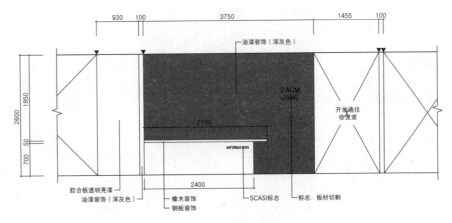

立面图C

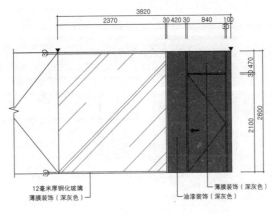

立面图D

■ 自助餐厅

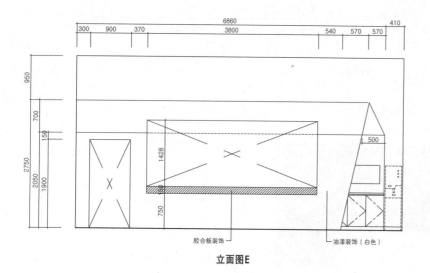

立面图E

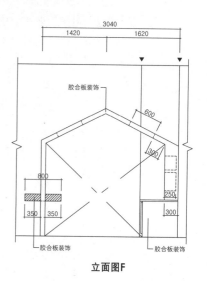

立面图F

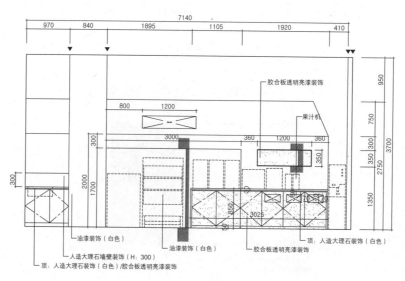

立面图G

游戏室

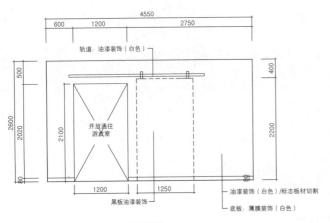

立面图H

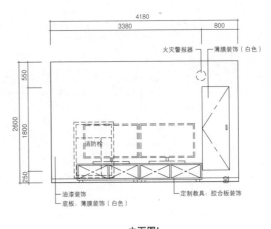

立面图I

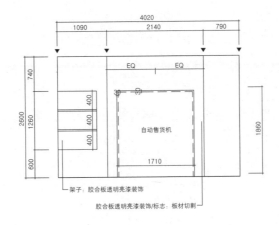

立面图J

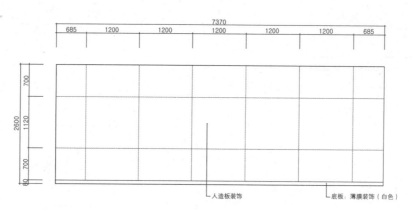

立面图K

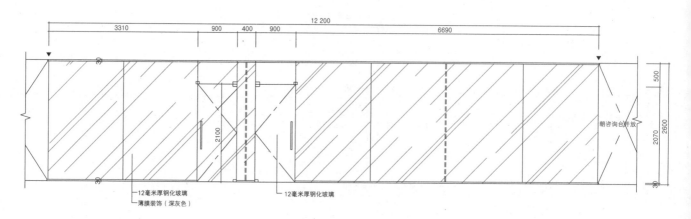

立面图L

会议室

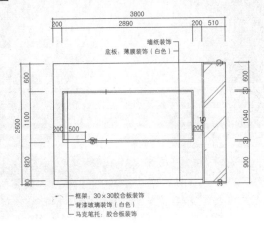

立面图M

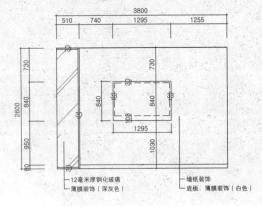

立面图N

大会议室

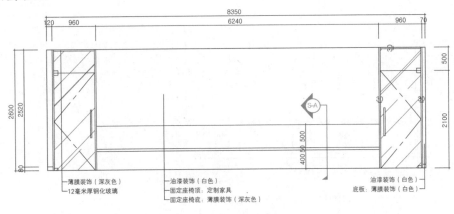

立面图O

剖面图A

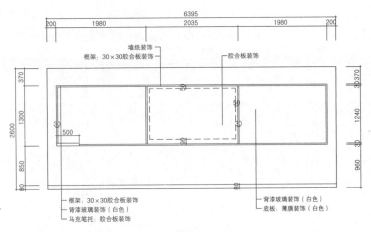

立面图P

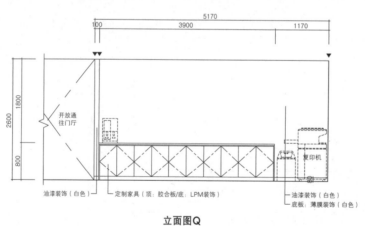

立面图Q

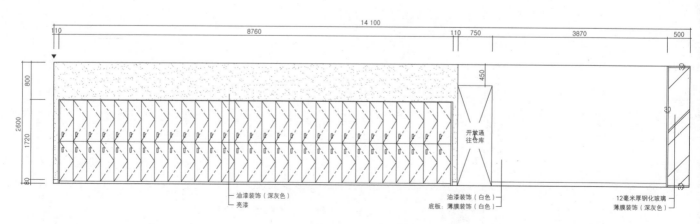

立面图R

74

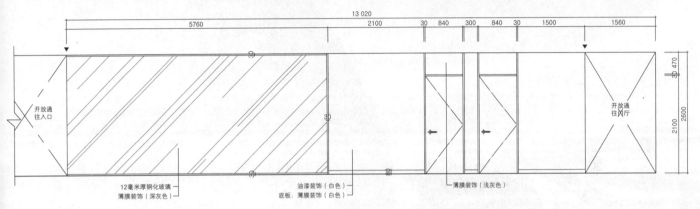

立面图S

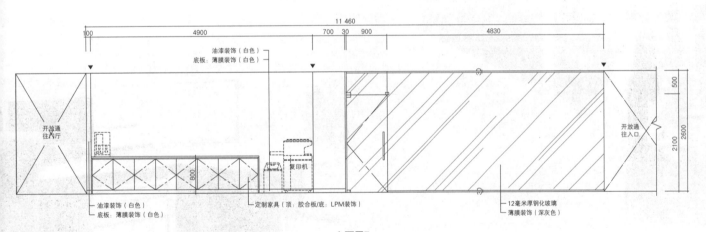

立面图T

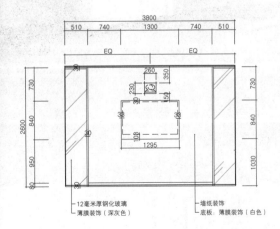

立面图U

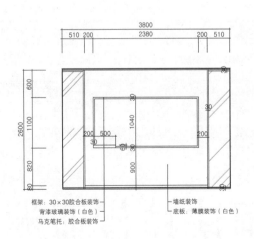

立面图V

Daum Games办公场所 | 75

阿法拉伐亚洲综合办事处

建筑面积：1390平方米

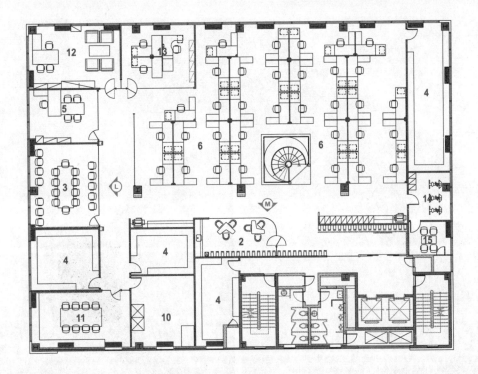

四层平面图

1. 入口
2. 接待室
3. 会议室
4. 仓库
5. 经理室
6. 办公室
7. 交流区
8. 团队工作室
9. 女休息室
10. 服务器机房
11. 培训室
12. 常务董事办公室
13. 财务室
14. 健身房
15. 吸烟区

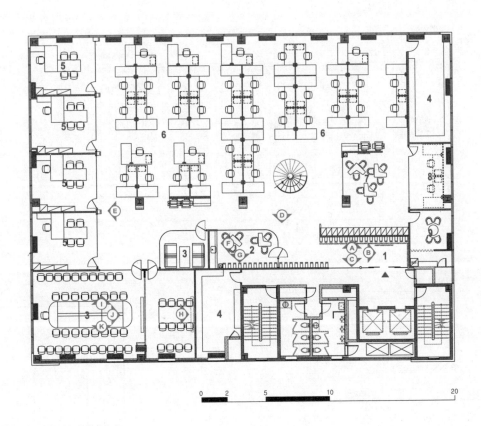

三层平面图

大厅

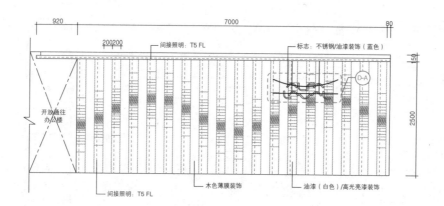

立面图A

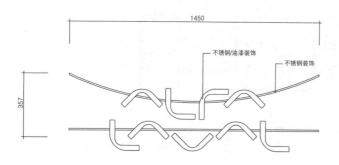

细节图A

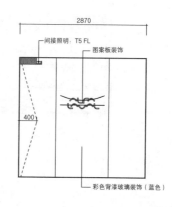

立面图B

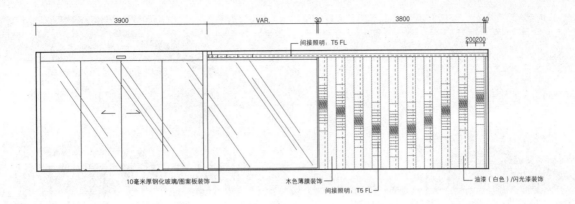

立面图C

形象墙细节图

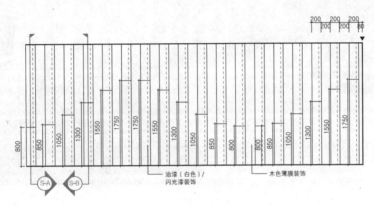

正立面图

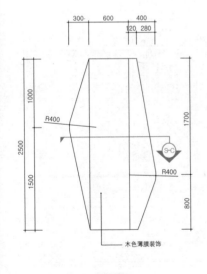

剖面图A

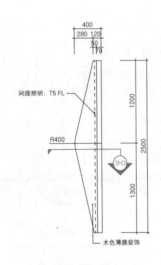

剖面图B

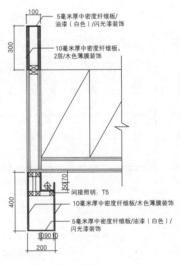

剖面图C

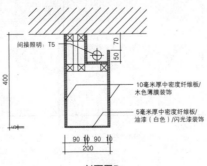

剖面图D

三楼办公室

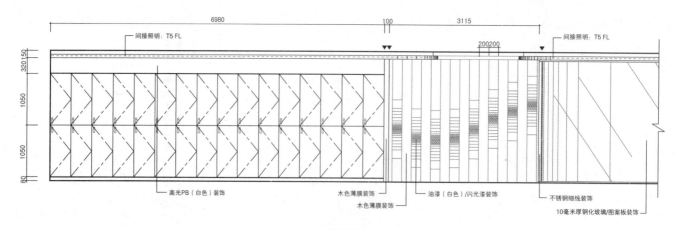

立面图D

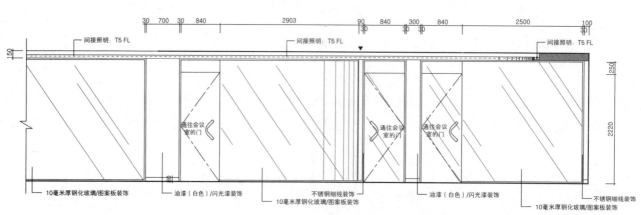

立面图D'

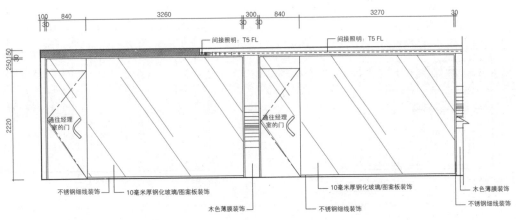

立面图E

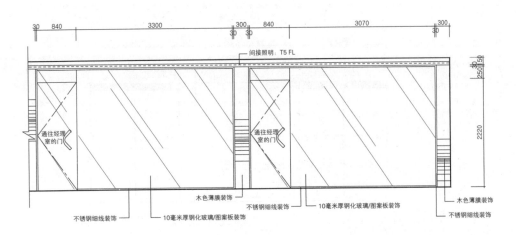

立面图E'

阿法拉伐亚洲综合办事处 **81**

■ 接待室

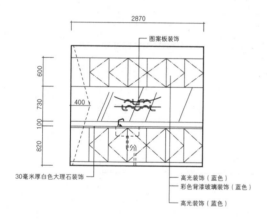

立面图F

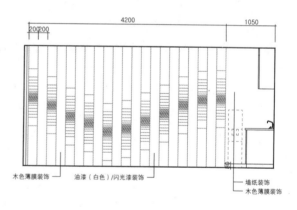

立面图G

■ 会议室

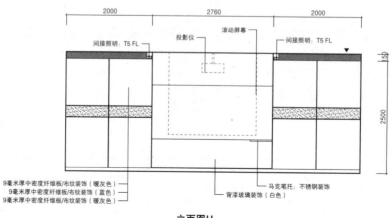

立面图H

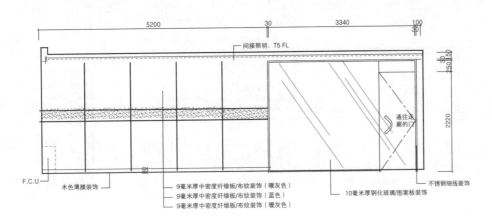

立面图I

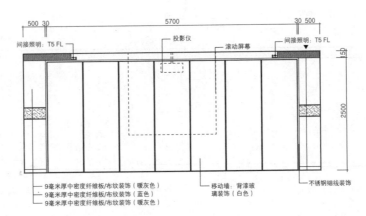

立面图J

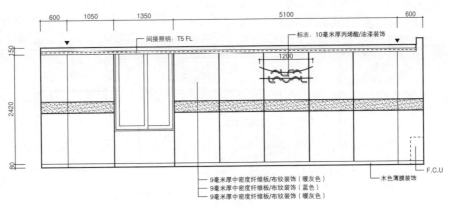

立面图K

四楼办公室

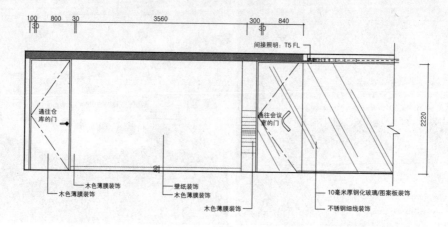

立面图L

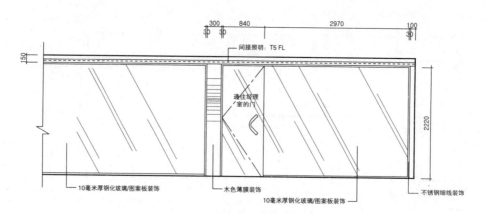

立面图L'

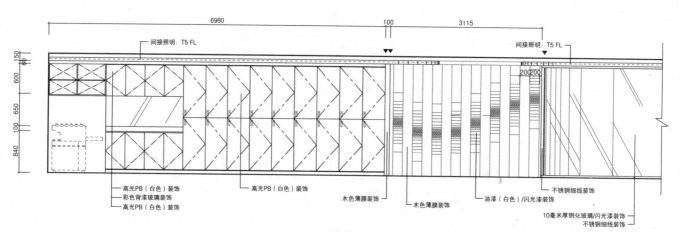

立面图M

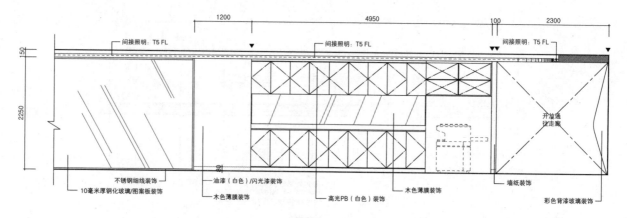

立面图M'

This 'Communication and Unity' as a concept connected the two floors of the office is to be done for naturally smooth work there. Generally, the use of white and blue tone provided a smart and trendy atmosphere. Blue Boomerang, ALFA LAVAL's CI imitated from the color and shape is graphicalized and applies to furniture, finishes and lighting to create a lively business environment.

At the entrance to the third floor, the inline boomerang-shaped three-dimensional structure is eye-caught and the design like wind-blowing brings some new energy into space. The image of the structure is used at the pillar and wall partitions as well as image wall to provide unity feelings. In addition, a point in space like the circular staircase is located inside there. This furnishes the feeling of rising-up to highlight the image of the company at the same time and meet inter-office communication consistent with the functional aspects.

"交流与统一"是办公室两层相衔接的概念,以便于自然流畅地工作。总体而言,白色和蓝色基调的使用,可以营造出一种醒目、时尚的氛围。阿法拉伐公司的企业标识——蓝色回旋镖,应用到家具、装饰和灯具上,营造一种活泼的企业环境。

第三层入口处,三维构造的回旋镖引人注目,风吹式的设计为该场所注入了一些新的活力。该构造的形象用在柱子、隔墙和图片墙上,给人一种统一的感觉。此外,圆形楼梯布置在其中,给人一种上升的感觉,凸显了公司的形象,同时在功能方面满足办公室间的交流。

| SMART MEDIA创新中心 |

建筑面积：264平方米

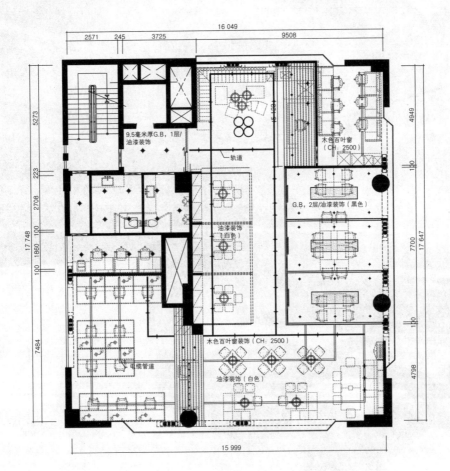

天花图

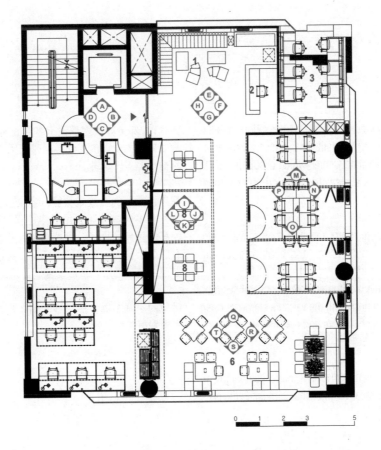

1. 休息区
2. 咨询台
3. 办公室
4. 演讲和大会议室
5. 书店咖啡馆
6. 开发空间
7. 咖啡馆
8. 会议室

平面图

Today, most of the offices boast various facilities for relaxation, meditation, self improvement, and hobbies bring vitality to the possibly monotonous work space. Smart Media Innovation Center, which occupies the 8th floor of the headquarters, provides a pleasant working environment reflecting such a trend.

The interior includes an open lecture room, meeting rooms, test beds, group project room, one-person project room, and book cafe for app development, technical support, and developer training. The natural monotone is accentuated by brilliant color to generate a sense of security. Each room is glazed to give out openness, and the variable elements open or divide the space according to the characters and approaches of works, thus maximizing the spatial use.

The optimized spatial composition offering diverse experiences heightens the work efficiency and represents a creative office that stimulates the passion.

如今，大部分办公室都设有各种设施供员工放松、冥想、自我提高，兴趣爱好能为单调的工作场所带来活力。该项目紧随时代潮流，提供一种愉悦的工作环境。

室内包括一间开放式教室、会议室、试验台、团队项目室、个人项目室和书店咖啡馆，用于应用程序研发、技术支持和研发者培训。明亮的颜色凸显了自然的单调，营造一种安全感。所有房间都安装有玻璃，给人一种宽敞的感觉，并按工作的特点和方式用各种元素展现或划分场所，从而使空间使用最大化。

优化的空间布局，提供不同的体验，有助于提高工作效率，同时也能激发员工的热情。

电梯厅

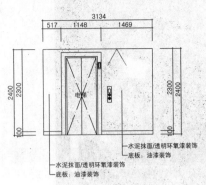

立面图A

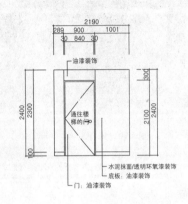

立面图B

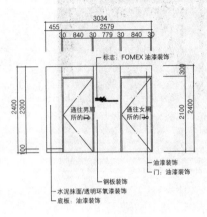

立面图C

立面图D

大厅休息区

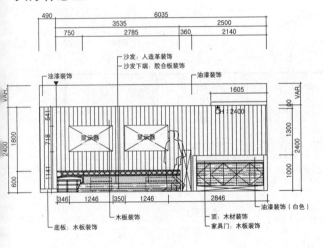

立面图E

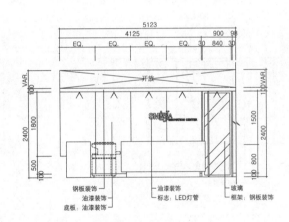

立面图F

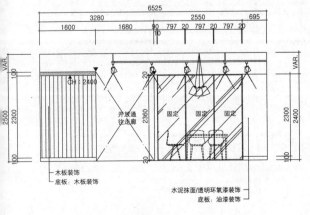

立面图G

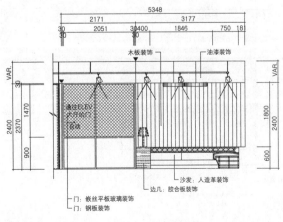

立面图H

SMART MEDIA 创新中心

会议室

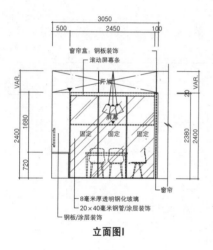

立面图I

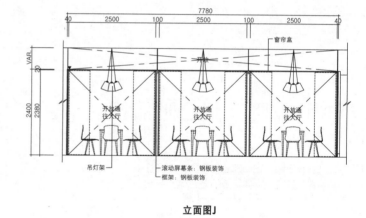

立面图J

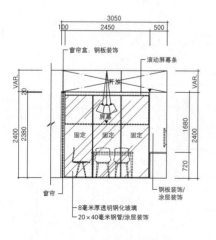

立面图K

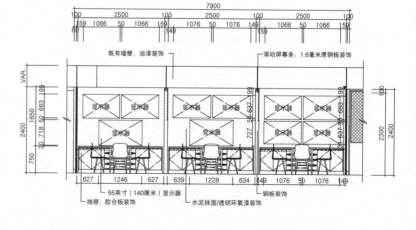

立面图L

演讲和会议室

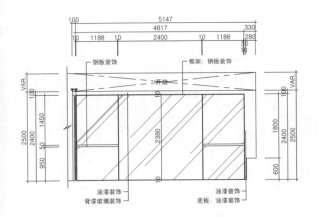

立面图M

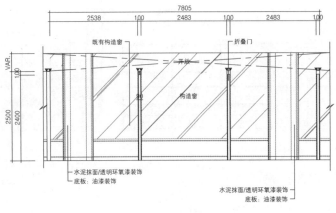

立面图N

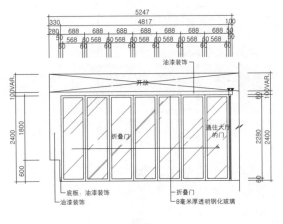

立面图O

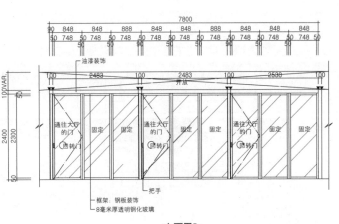

立面图P

SMART MEDIA 创新中心

■ 研发空间

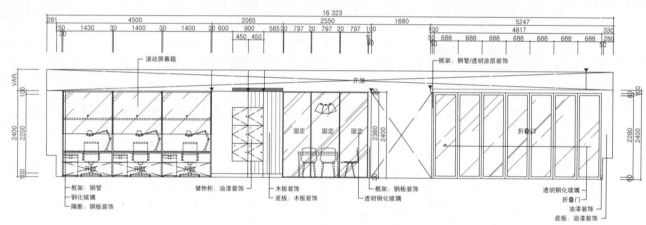

立面图Q

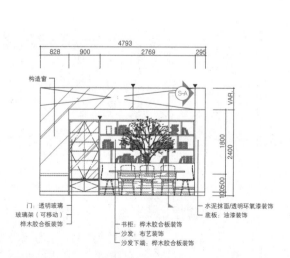

立面图R

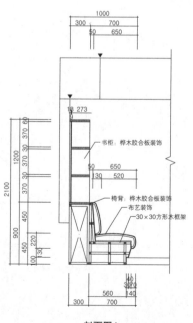

剖面图A

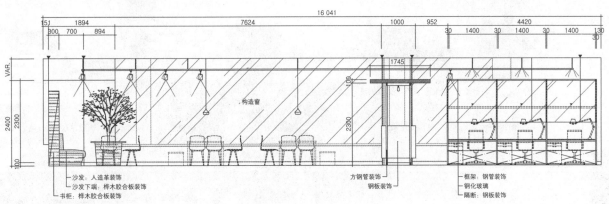

立面图S

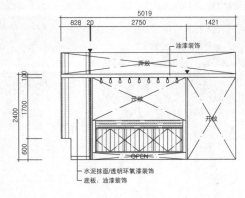

立面图T

| Kkotsbom办公场所 |

建筑面积：316平方米

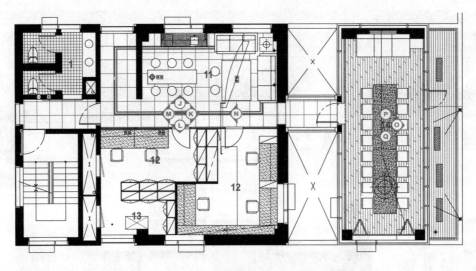

三层平面图

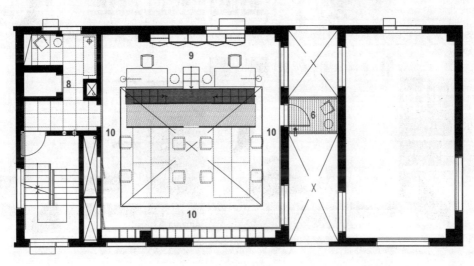

阁楼平面图

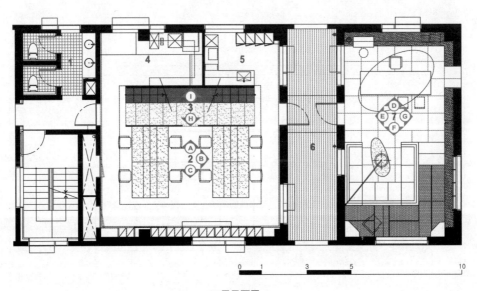

二层平面图

1. 厕所
2. 设计团队办公室
3. 楼梯
4. 影印室
5. 餐厅
6. 露天平台
7. 总裁办公室
8. 图书室
9. 团队经理办公室
10. 走廊
11. 接待室
12. 办公室
13. 资料室
14. 会议室

■ 设计团队办公室

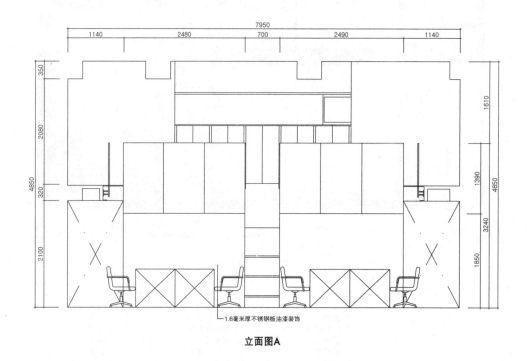

立面图A

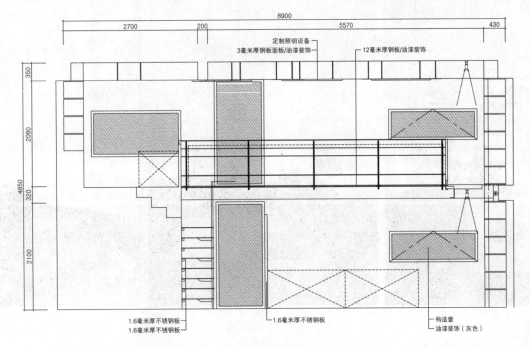

立面图B

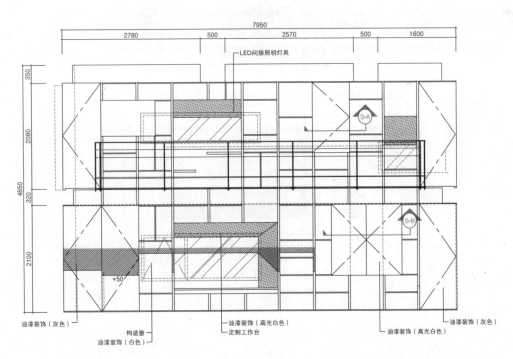

立面图C

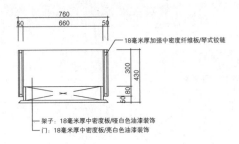

剖面图A

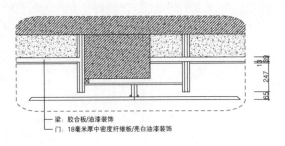

剖面图B

■ CEO办公室

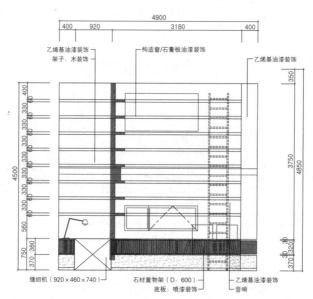

立面图D

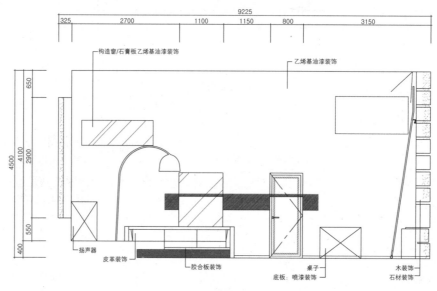

立面图E

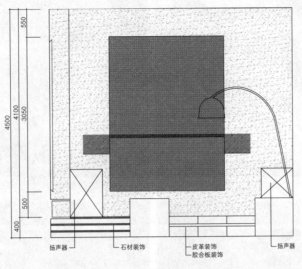

立面图F

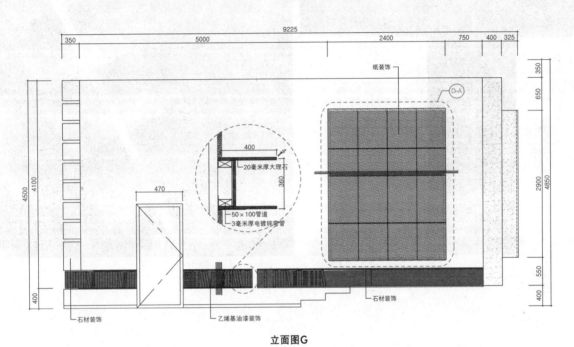

立面图G

细节图A

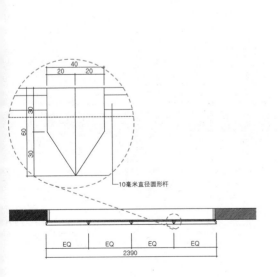

俯视图

正立面图

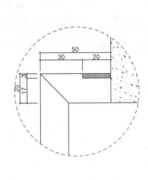

剖面图C

■ 楼梯

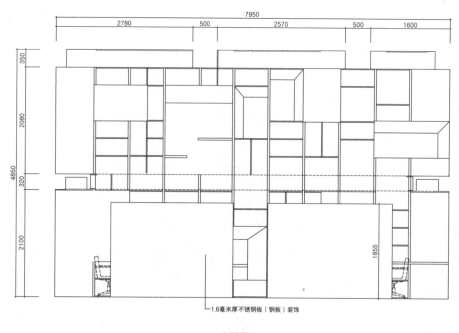

立面图H

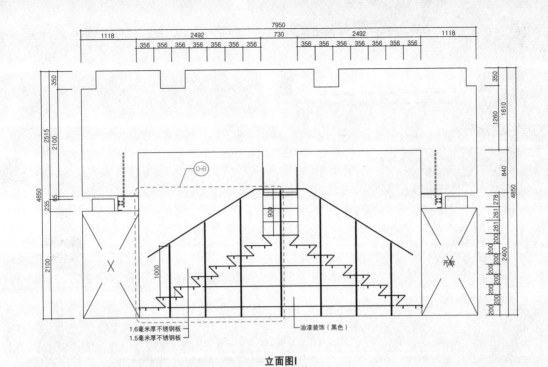

立面图I

细节图B

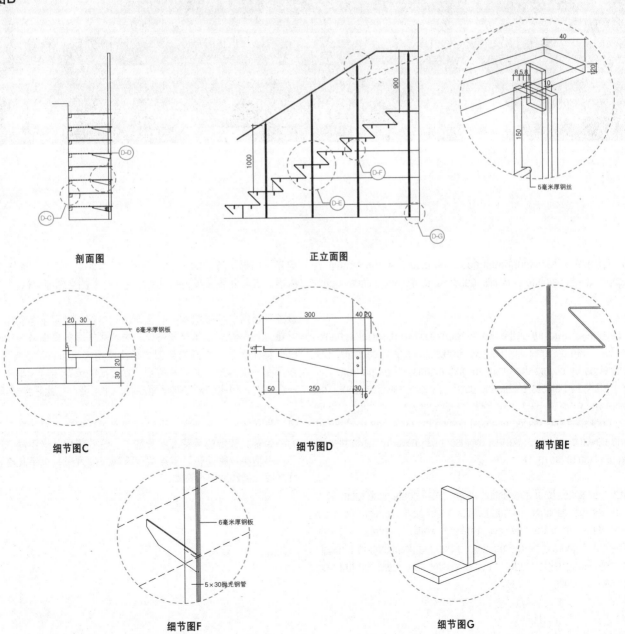

剖面图　　　　正立面图

细节图C　　　细节图D　　　细节图E

细节图F　　　细节图G

Film poster maker 'Kkotsbom' was designed based on the motive of 'Chaekgado', a folk painting of books, potteries, stationery, incense burners.

In the mezzanine structure of the design room are five-meter-high bookshelves filled with design-related books, workstations, and benches. The white panel partly covers the desks to give a sense of proportion. The bridges installed in places enable the users to move, read, or work. The president's room on the first floor took the advantage of its high ceiling height; a bookshelf was placed to a side behind the desk, and the floor is lifted and covered by tatami so that the users can have a seat to have a meeting or listen to the music.

The third floor consists of a reception room, small conference room, and the offices for the management support team and planning team. Its cozy ambience comes from the low ceiling height, which is unlike the lower floor. The small conference room shows a humble combination through the six-meter-long table made of epoxy-coated linen and the lighting made of Korean paper.

电影海报制作商"Kkotsbom"是根据"Chaekgado"（一种与绘画图书、陶器、文具和香薰有关的民间传统文化）的主题而设计的。

该办公场所的楼层结构中有五米高的书架，放置了各种与设计相关书籍，且布置有工作台和长椅。白色面板覆盖桌面的一部分，营造出一种比例感。空间内架起的桥让使用者可以移动、阅读或工作。布置在一层的总裁办公室利用了其较高的天花高度，桌子后侧设置一个书架，用榻榻米提高地面高度，使用者可在此开会或听音乐。

第三层包括一间接待室、小型会议室以及管理支持团队和规划团队的办公室。较低的天花高度营造了舒适的氛围，这不同于下面的楼层。小型会议室通过环氧涂层亚麻制成的六米长桌子和纸制的照明灯具展现简约的组合风格。

■ 接待室

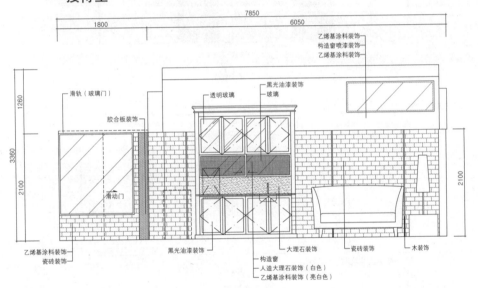

立面图J

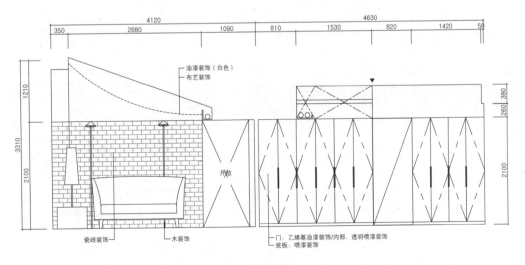

立面图K

■ 走廊

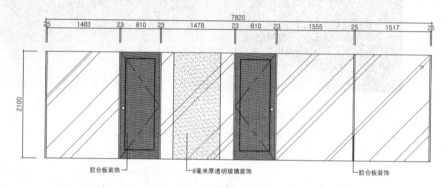

立面图L

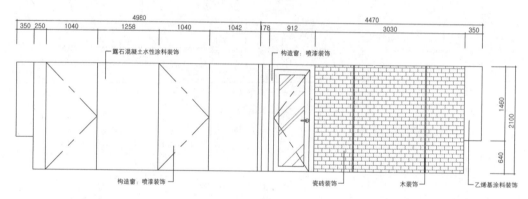

立面图M

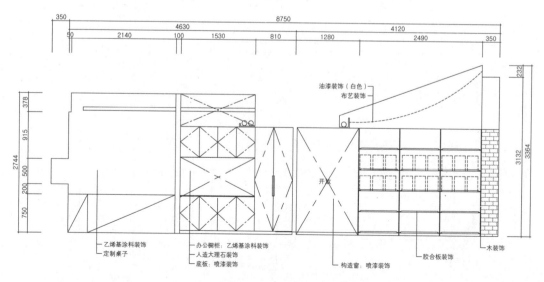

立面图N

会议室

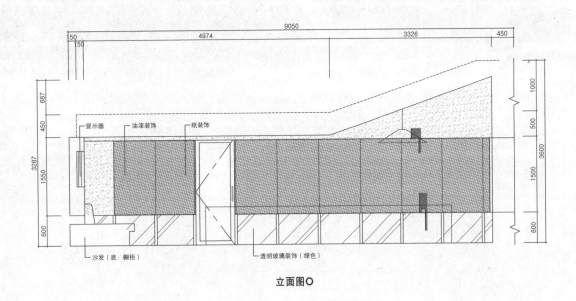

立面图 O

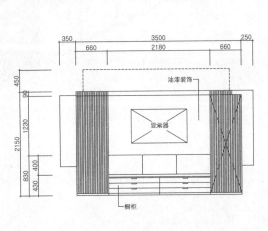

立面图 P

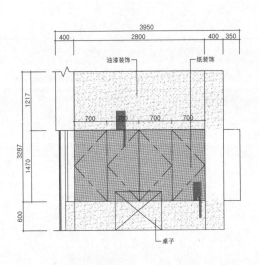

立面图 Q

Riot Games 办公场所

建筑面积：896平方米

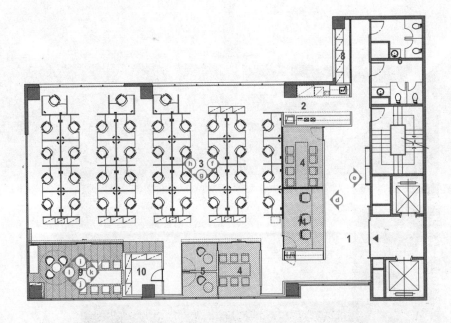

十四层平面图

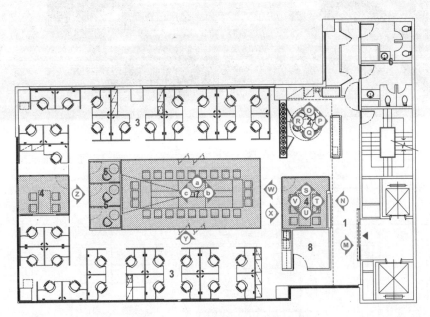

九层平面图

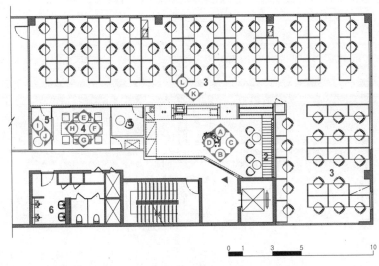

三层平面图

1. 大厅
2. 休息区
3. 办公室
4. 会议室
5. 电话间
6. 厕所
7. 大会议室
8. 仓库
9. 总裁办公室
10. 档案室
11. 机房

Riot Games 办公场所　107

■ 三层大厅

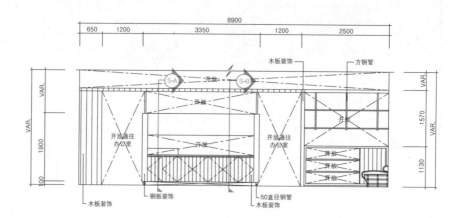

立面图A

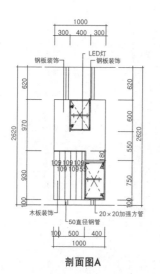

剖面图A

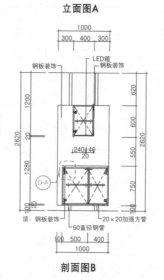

剖面图B

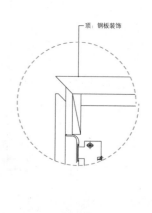

细节图A

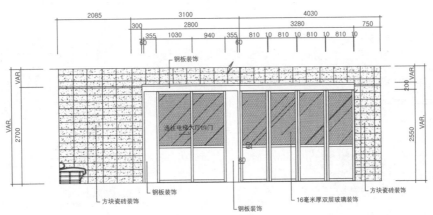

立面图B

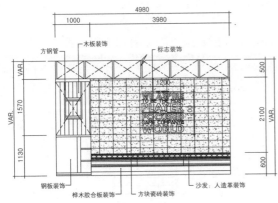

立面图C

立面图D

Riot Games 办公场所

三层会议室

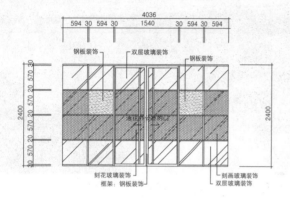

立面图E　　　　　　　　　立面图F

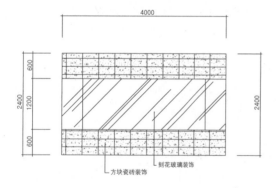
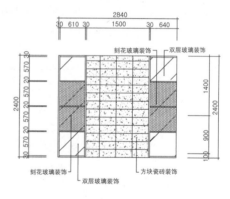

立面图G　　　　　　　　　立面图H

电话间

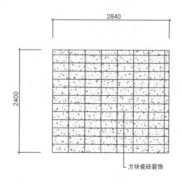
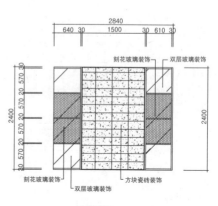

立面图I　　　　　　　　　立面图J

三层办公室

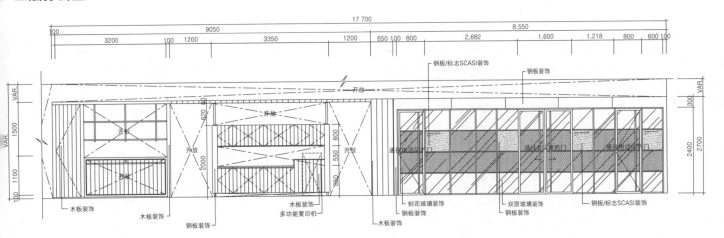

立面图K

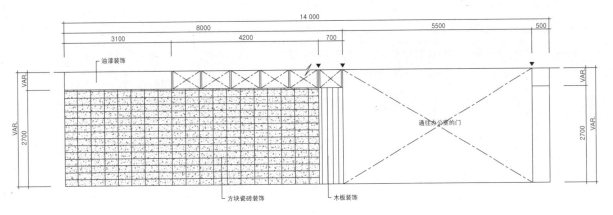

立面图L

九层大厅

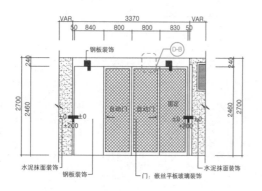

立面图M

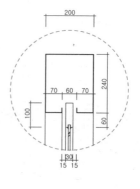

细节图B

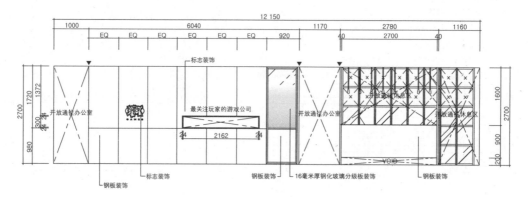

立面图N

九层休息区

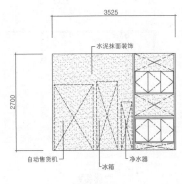

立面图O

立面图P

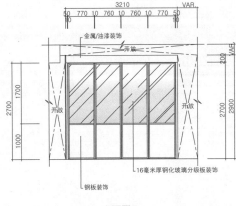

立面图Q

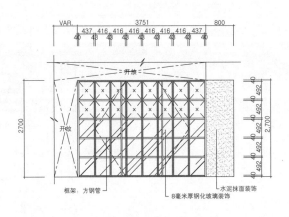

立面图R

九层会议室

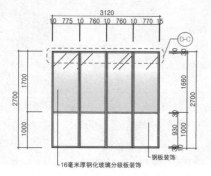

立面图S

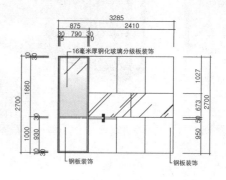

立面图T

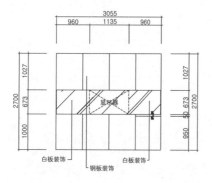

立面图U

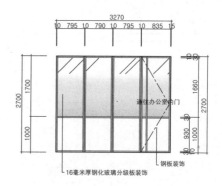

立面图V

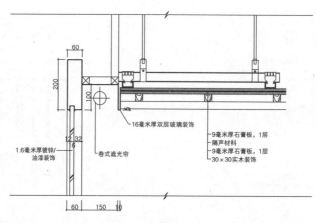

细节图C

The Riot Games, game development company based in California, the United States. This company consists of professional staffs for web management, design, and hardware and requires working space considering the convenience of the employees since they have to be seated for long hours.

Offices divided into three floors have exposed ceiling so that it could visually expand the space and used contrasting materials; heavy and cool materials such as metal, glass, cement, and toned down colors in one hand and warm materials like wood louvers on the other to make the space more stable and pleasant.

Resting area in the center of the entrance of the third floor has been opened since it has to accommodate a large group of people in a relatively limited space and added open-shaped closets, cabinets and neon lights to mitigate stuffiness. The office leading directly to it has been decorated with bright and natural material to express comfortable atmosphere and has local illumination so that employees can adjust the level of illumination and relieve the stress of eyes. On the 9th floor, there is a glass box-shaped lager conference room, small conference room and resting room at the center and arranged seats for working at the three corner of the place. In particular, the grid-shaped wall in the resting space has a large flowerpot inside so that it could separate the space from the working place and express the pleasant atmosphere at the same time. The 14th floor, where there is the office for president, is able to give out strong impression since it has elongated sign wall from the entrance and has cozy atmosphere as if it were a cafe with awining, old brick and patterned bars.

This place completes a pleasant atmosphere by efficiently allocating spaces according to the function of each space and using natural materials as objets which can offset stuffy feeling that a working space can have.

Riot Games 是美国加利福尼亚州的一家游戏开发公司。该公司由网络管理、设计和硬件几方面的专业人员组成，他们需要长时间伏案工作，所以工作场所需考虑员工的便利。

分布在三个楼层的办公室均采用开放式天花板，这样可以从视觉上拓展空间，并运用了对比材料；一方面采用金属、玻璃、水泥等冷重的材料和统一柔和的颜色，另一方面采用木百叶窗等温暖的材料，让办公场所更加安定愉悦。

第三层入口中间的休息区已经开放，在相对有限的空间可以容纳大量人群，另外增添了开放式壁橱、橱柜和霓虹灯来缓解沉闷感。直接通向休息区的办公室用明亮自然的材料装修，营造出舒适的氛围，并设置局部照明，这样员工可以调节亮度，缓解眼部疲劳。第九层中间设有玻璃箱式的大型会议室、小型会议室和休息区，并在三个角落设置工作座椅。休息区网格墙内布置有大花盆，把工作区与休息区区分开来，同时营造出愉悦的氛围。第十四层设有总裁办公室，从入口开始有一堵细长的标志墙，如同拥有复古式砖头和刻花钢筋的咖啡厅，有着舒适的氛围，给人留下深刻的印象。

根据各区域的功能有效分布空间，采用能缓解工作场所沉闷感的自然材料，营造一种愉悦的氛围。

■ 走廊

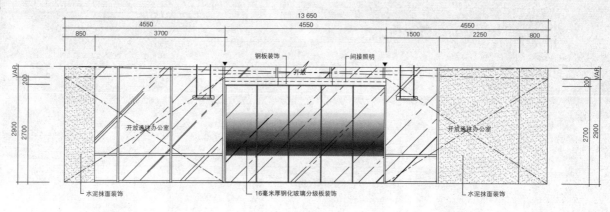

立面图W

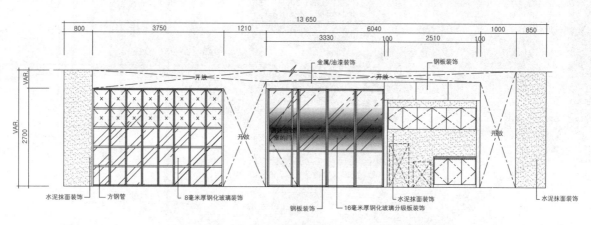

立面图X

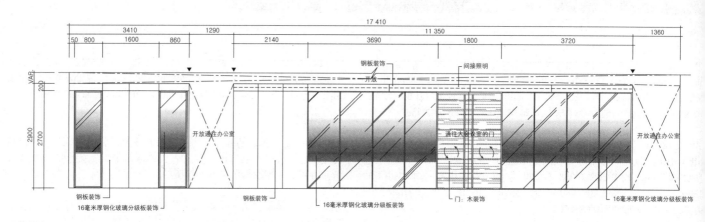

立面图Y

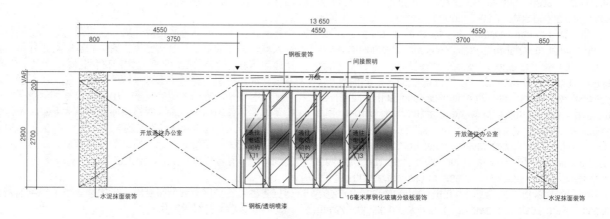

立面图Z

大会议室

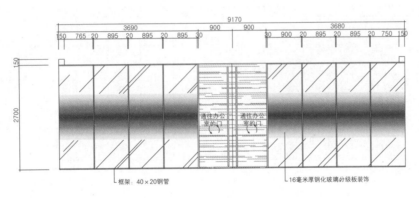

立面图a

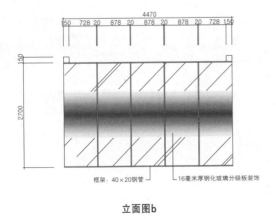

立面图b

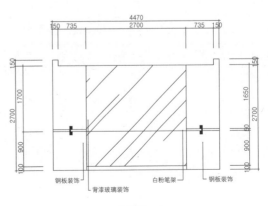

立面图c

■ 十四层大厅

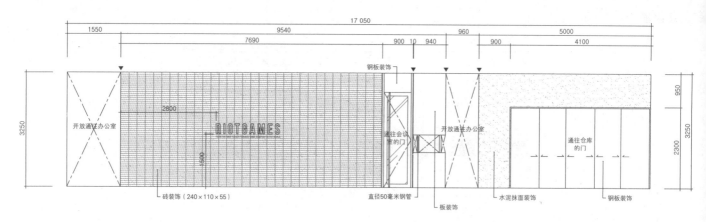

立面图d

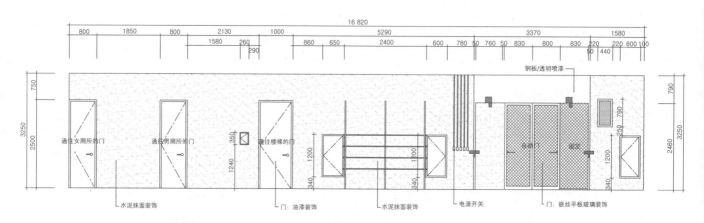

立面图e

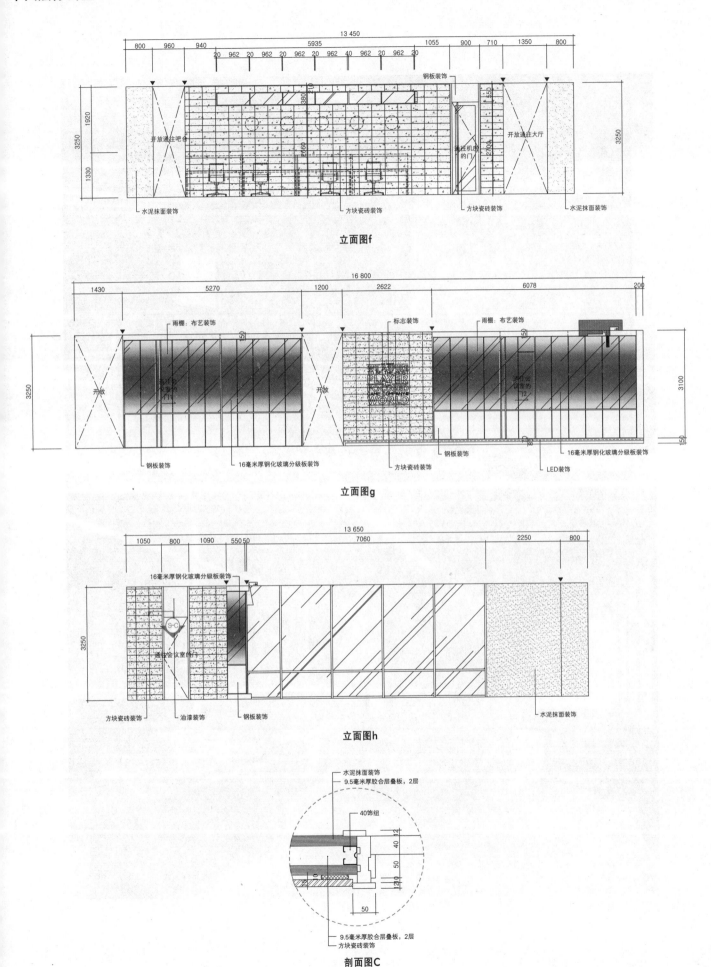

总裁办公室

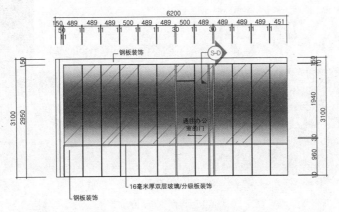

立面图 i

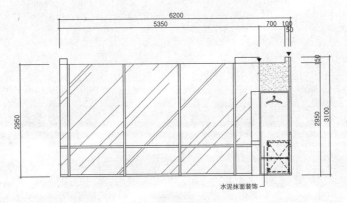

立面图 j

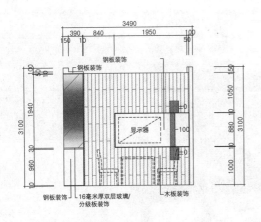

立面图 k

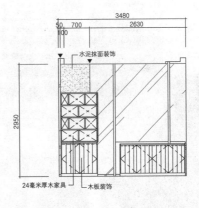

立面图 l

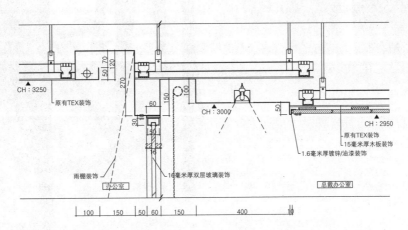

剖面图 D

| Tapjoy办公场所 |

建筑面积：314平方米

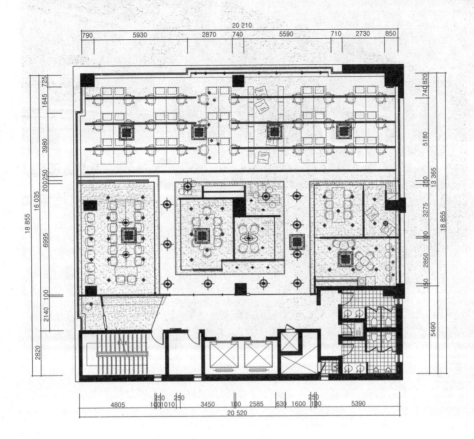

天花图

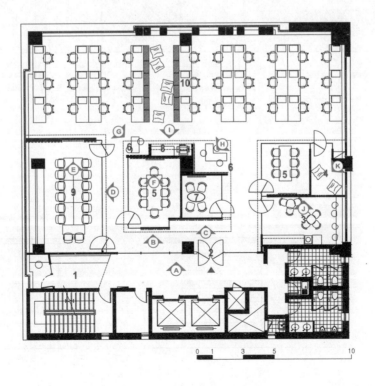

1. 电话区
2. 入口
3. 配餐室
4. 休息区
5. 会议室
6. 自由会见区
7. 接待室
8. 影印室 & 餐厅
9. 大会议室
10. 办公室

平面图

Tapjoy办公场所

■ 入口

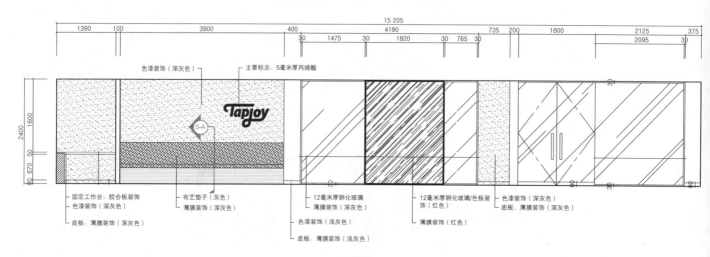

立面图A

■ 形象墙

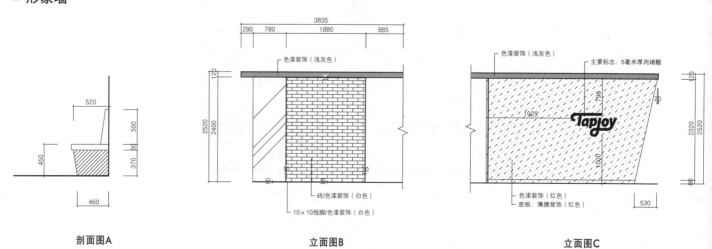

剖面图A　　　　　　立面图B　　　　　　立面图C

走廊

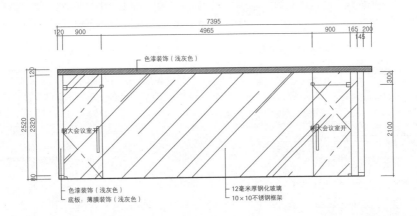

立面图D

大会议室

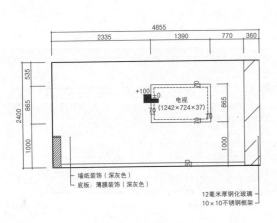

立面图E

会议室

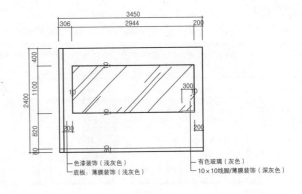

立面图F

Tapjoy 办公场所

■ 自由会见区

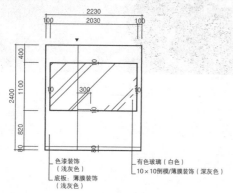

立面图G

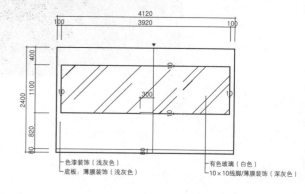

立面图H

■ 影印室和餐厅

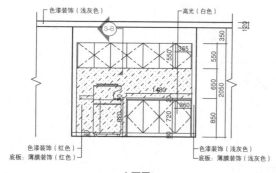

立面图I

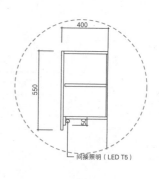

剖面图B

■ 配餐室

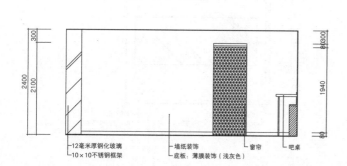

立面图J

■ 休息区

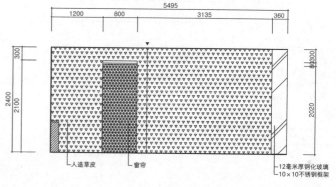

立面图K

Mobile advertising firm 'Tapjoy' has opened a new creative office that breaks away from the working environment usually found.

Since most of the works are done in the frame work of a project, the team members change accordingly, thus calling for a design that focuses on communication between individual employees. The overall ceiling was exposed to deliver a sense of openness, while the board wall and bar table were placed to support informal meetings to take place. The designer also introduced the seating type working and relaxing area covered with artificial turf in the middle of the office, the individual phone booth for employees, and the linear lighting that reduces the fatigue of the eyes for convenience. The rough texture of the black steel plates and Oregon pine veneer used on the wall and furniture give a natural look, while the fire protection piping beneath the exposed ceiling was painted in red to enliven the office along with the image wall at the entrance.

手机广告公司 Tapjoy 新增了一处不同于常见的办公环境的创意办公场所。

由于大部分工作是在项目的框架下完成，团队成员会发生变化，因此设计需要强调员工之间的交流。所有天花板外露，传递一种开放的感觉，同时设置板墙和吧桌，可供进行非正式会议。设计师还在办公室中间设置了工作和休息相结合的区域，上面覆盖人造草皮；为员工设置了私人电话间，还引进了线性灯光，以缓解眼睛疲劳。墙壁和家具所用的黑钢板和花旗松胶合板粗糙的质地呈现自然的外观，而将外露天花板下的防火管道系统涂上红漆，与入口处的形象墙共同活跃了整个办公空间的气氛。

教育机构

- DongEun-I幼儿园
- Presby教育中心
- KUT 环球教育中心
- 人寿保险

| DongEun-| 幼儿园 |

建筑面积：1494平方米

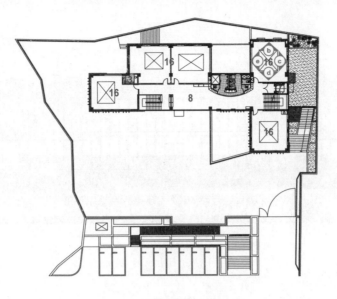

二层平面图

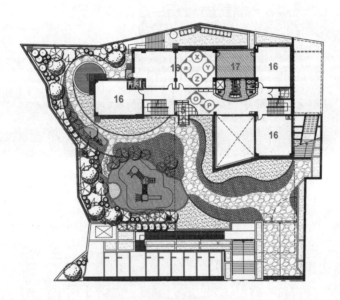

一层平面图

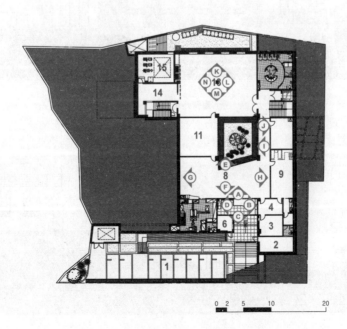

地下一层平面图

1. 停车场
2. 仓库
3. 套间
4. 校长办公室
5. 入口
6. 保安室
7. 厨房
8. 大厅
9. 教师办公室
10. 院子
11. 画室
12. 厕所
13. 体育教学室
14. 准备室
15. 机械房
16. 教室
17. 室外游戏区
18. 游泳池
19. 操场

DongEun-I 幼儿园

■ 外观

北侧外观

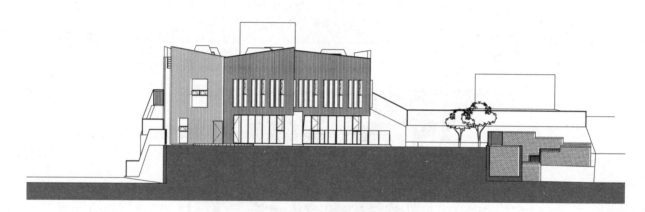

东侧外观

南侧外观

西侧外观

The design introduces the characters of origami, which opens, closes, and folds paper, thus creating a unique structure made of various forms. The classrooms, corridors, hall, supplementary area, and courtyard are separated by folding doors and moving walls. The central courtyard can become a field study site by extending the classrooms, while the exterior of the building can be used as a blackboard in an outdoor learning space. The ceiling of the second floor classroom soars up into the sky, and the glass and various colors finish the vibrant space.

Bringing the building itself in actual education through creative approach, DongEun-I Kindergarten offers an innovative educational space for children to run around, play and learn in nature.

设计引进了折纸的特点，通过打开、闭合、折叠打造由各种造型构成的独特结构。教室、走廊、大厅、辅助区和院子通过折叠门和移动墙分隔。延伸的教室可将中央的院子变成室外学习区，而在室外学习区，建筑的外表面可作为黑板。第二层教室的天花板很高，用玻璃并涂上各种颜色进行装饰。

DongEun-I 幼儿园通过创新的方法将自身的建筑融入实际教学中，为儿童在大自然中奔跑、玩耍和嬉戏提供了一处独特的学习场所。

■ 地下一层入口

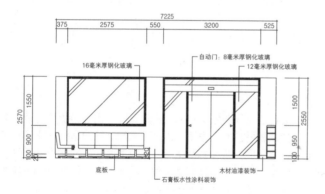

立面图A

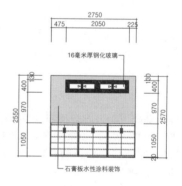

立面图B

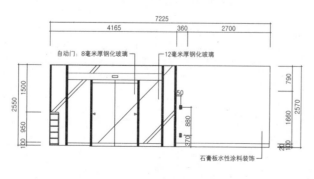

立面图C

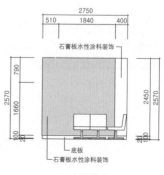

立面图D

入口大厅

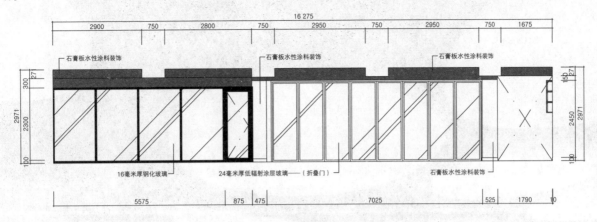

立面图E

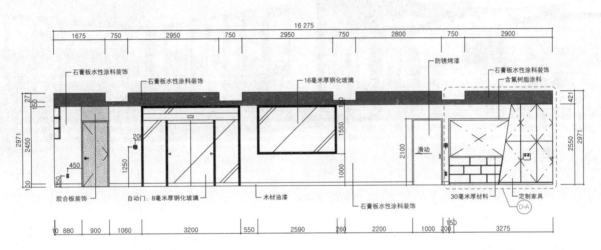

立面图F

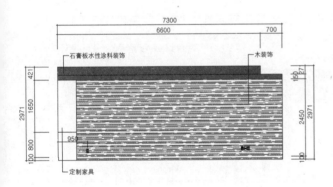

立面图G

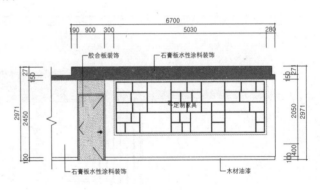

立面图H

细节图A——家具

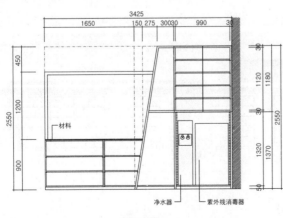

正立面图

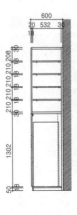

剖面图

地下一层走廊

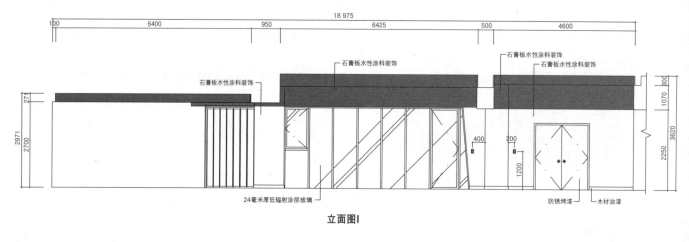

立面图 I

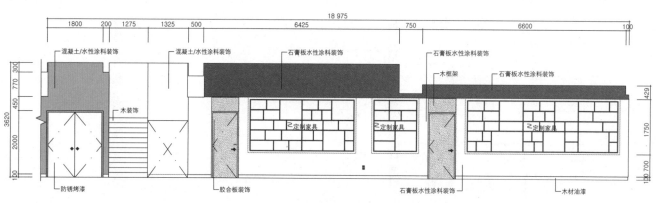

立面图 J

体育教学室

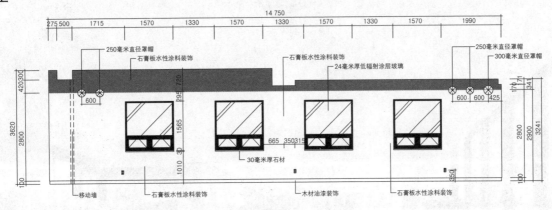

立面图 K

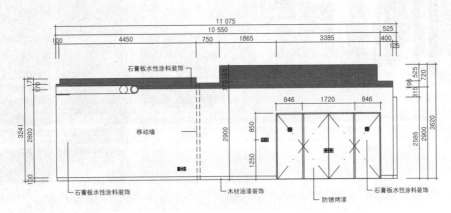

立面图 L

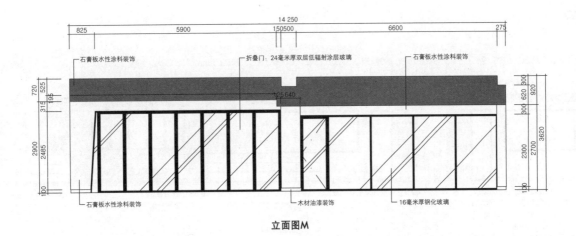

立面图 M

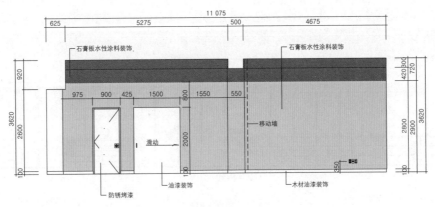

立面图 N

一楼走廊

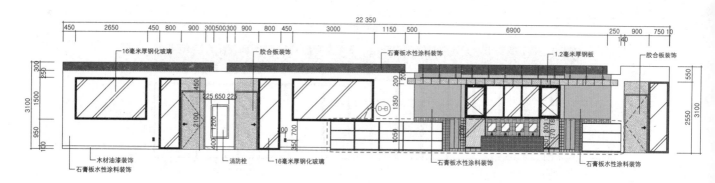

立面图O

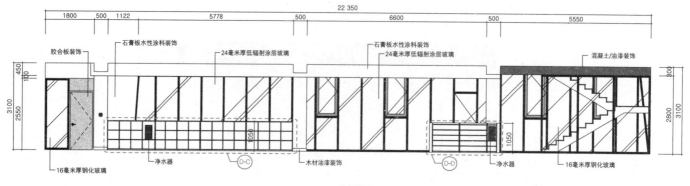

立面图P

■ 细节图B

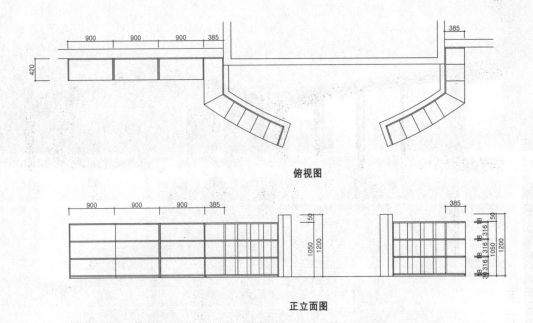

俯视图

正立面图

■ 细节图C

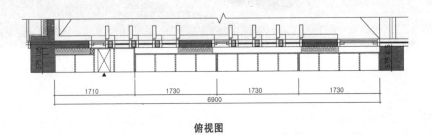

俯视图

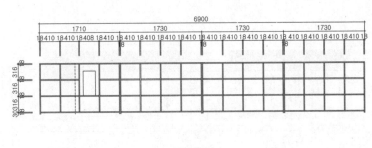 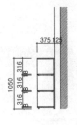

正立面图　　　　　　　　　剖面图

■ 细节图D——鞋架

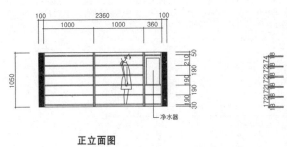 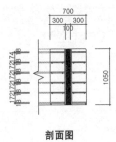 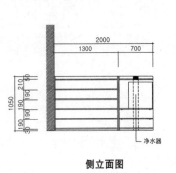

正立面图　　　　　　剖面图　　　　　　侧立面图

■ 一楼厕所

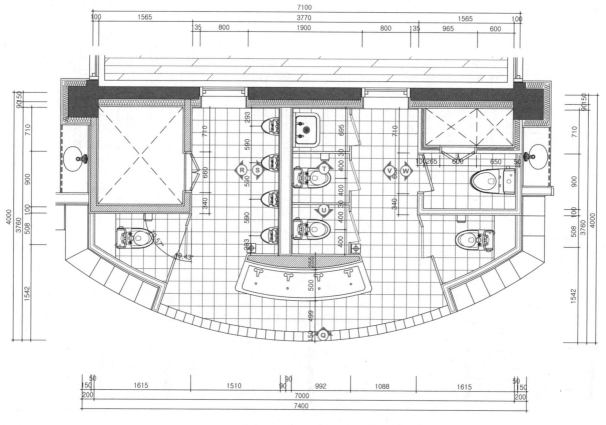

平面图

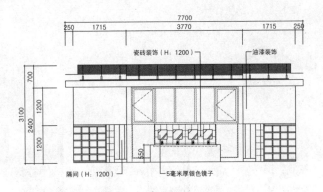

立面图Q

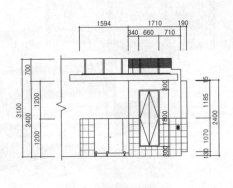

立面图R

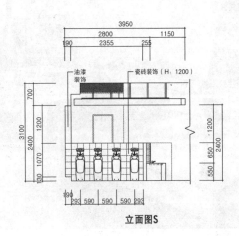

立面图S

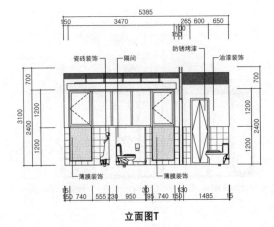

立面图T

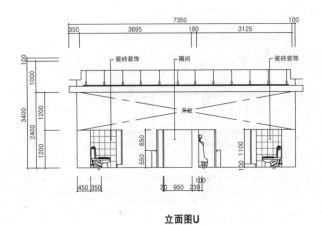

立面图U

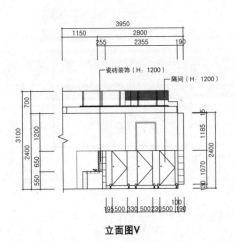

立面图V

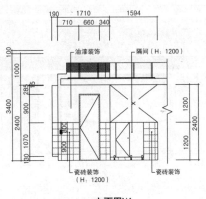

立面图W

一楼教室

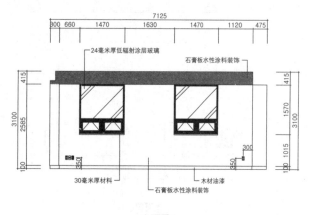

立面图X

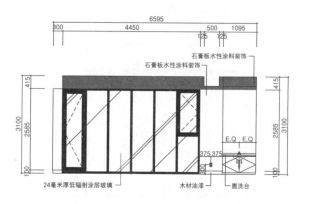

立面图Y

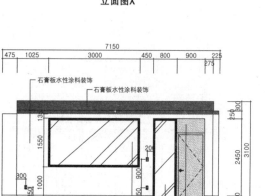

立面图Z

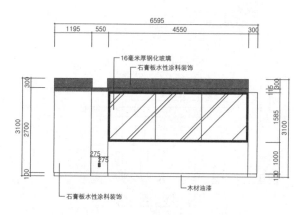

立面图a

二楼教室

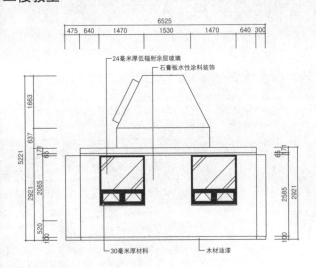

立面图b

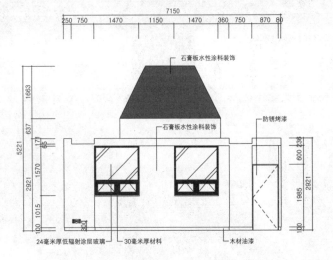

立面图c

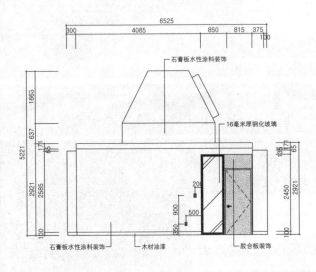

立面图d

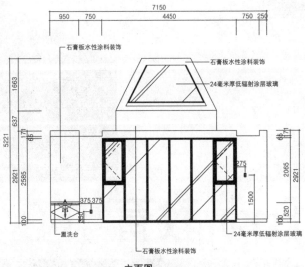

立面图e

Presby教育中心

建筑面积：2272平方米

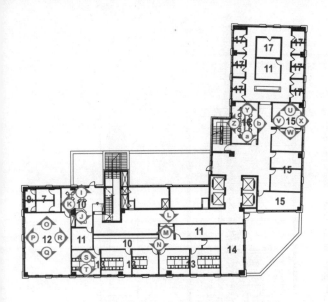

三层平面图

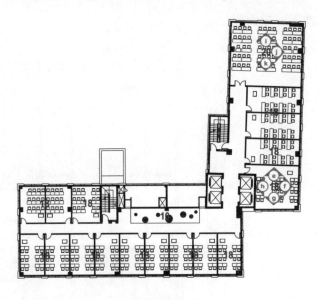

五层平面图

1. 停车场
2. 大厅
3. 教师办公室
4. 影印室
5. 咨询室
6. 主任办公室
7. 更衣室
8. 跆拳道室
9. 浴室
10. 露天场所
11. 教师等候室
12. 芭蕾舞排练室
13. 美术教室
14. 展览厅
15. 科学教室
16. 围棋室
17. 音乐教室
18. 教室
19. 图书室
20. 研讨室

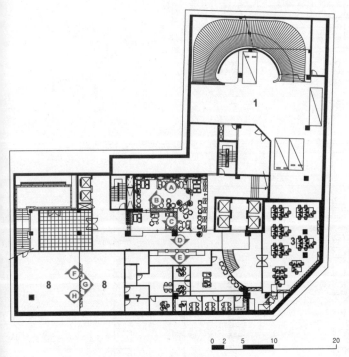

地下一层平面图

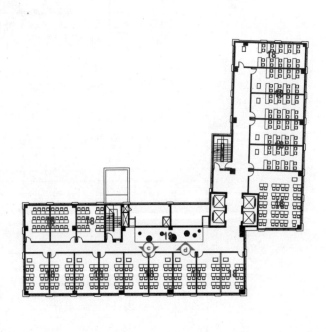

四层平面图

Presby 教育中心

地下一层大厅

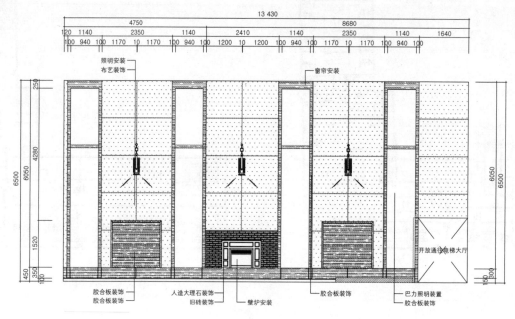

立面图A

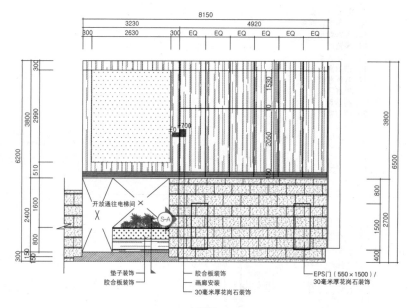

立面图B

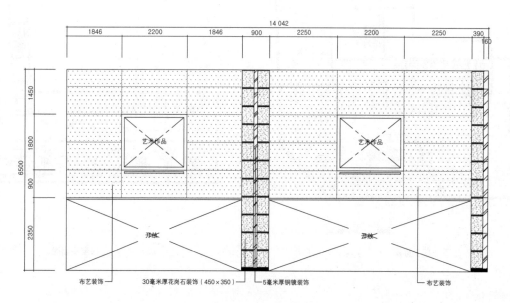

立面图C

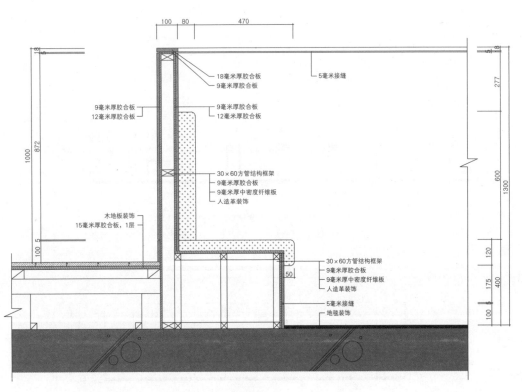

剖面图A

■ 接待室

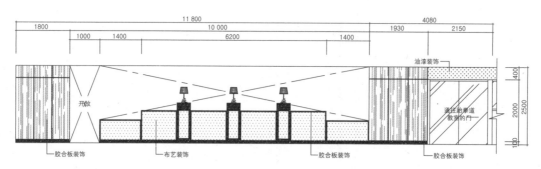

立面图D

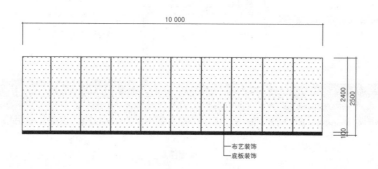

立面图E

跆拳道教室

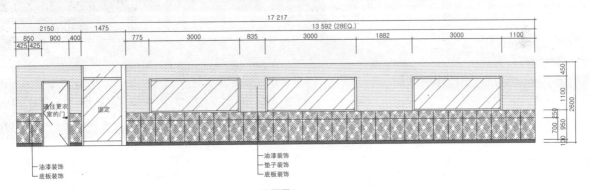

立面图F

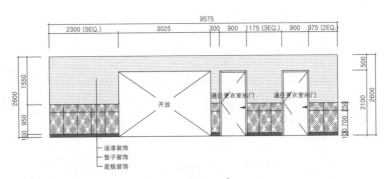

立面图G

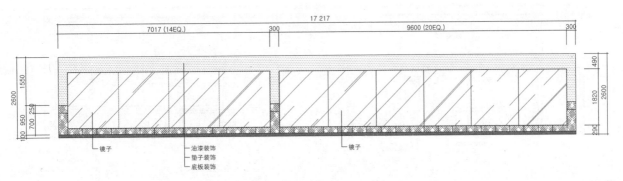

立面图H

三层开放空间

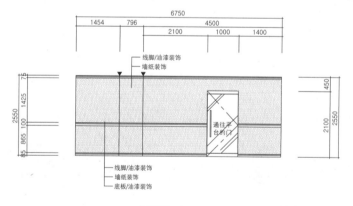

立面图I

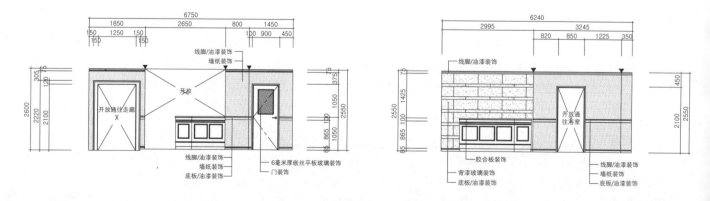

立面图J　　　　　　　　　立面图K

三层走廊

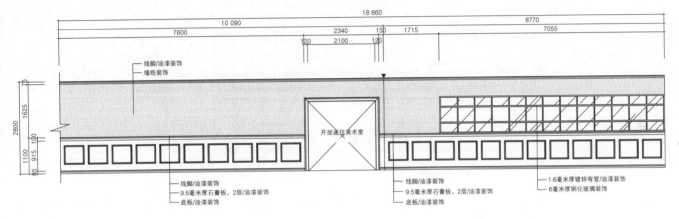

立面图 L

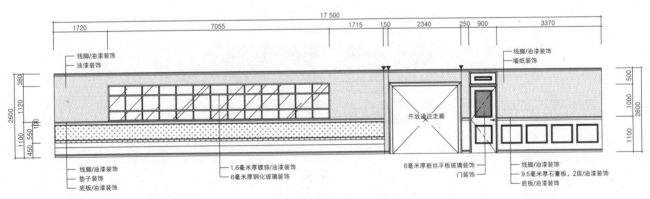

立面图 M

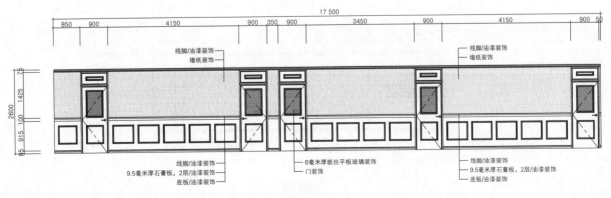

立面图 N

芭蕾舞排练室

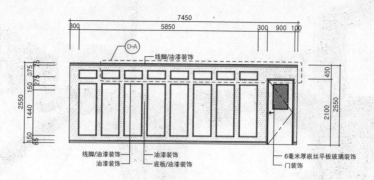

立面图O

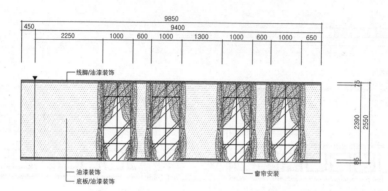

立面图P

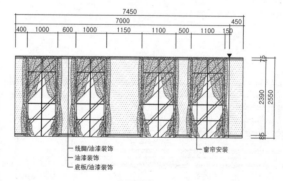

立面图Q

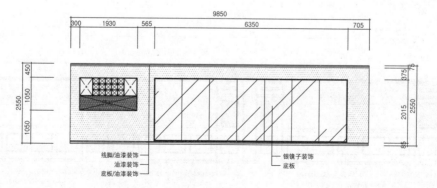

立面图R

芭蕾舞排练室天花板

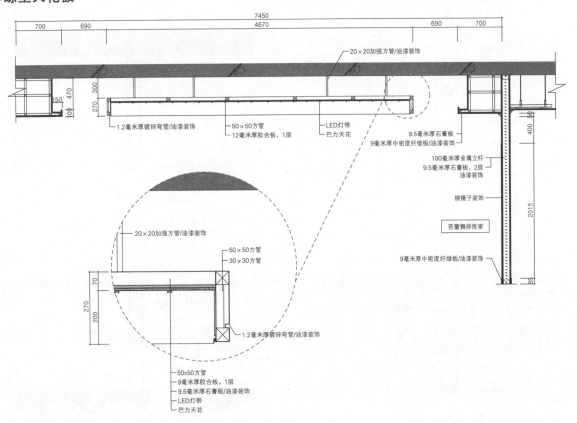

细节图A

美术教室

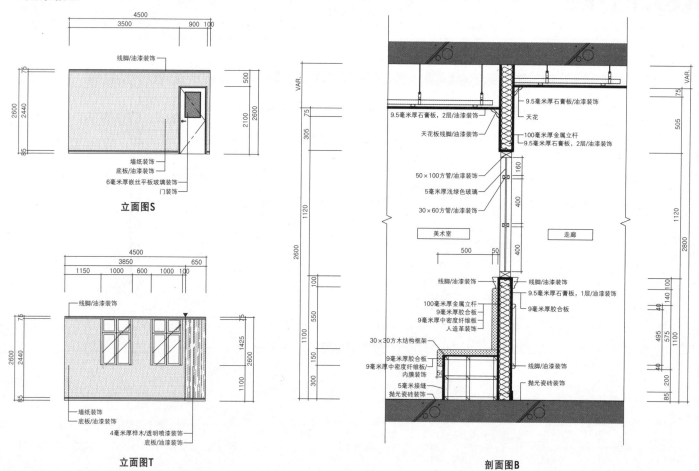

立面图S

立面图T

剖面图B

科学教室

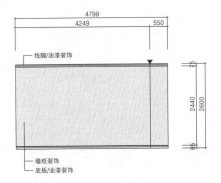

立面图U

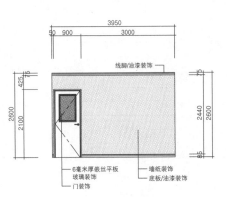

立面图V

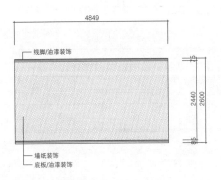

立面图W

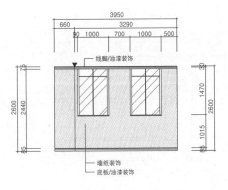

立面图X

Presby 教育中心

围棋室

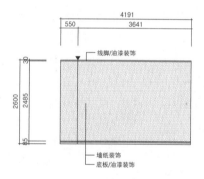

立面图Y

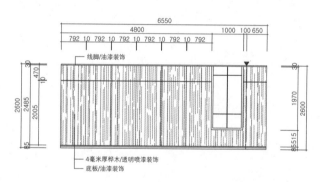

立面图Z

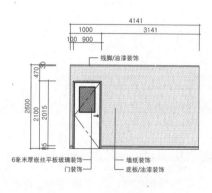

立面图a

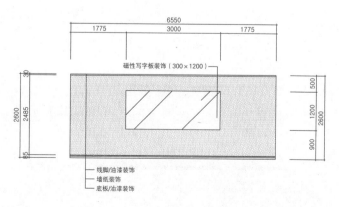

立面图b

■ 四层走廊

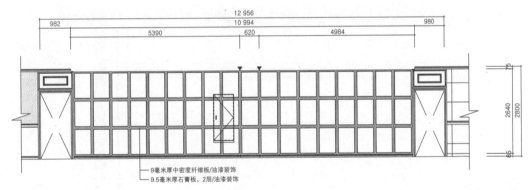

立面图c

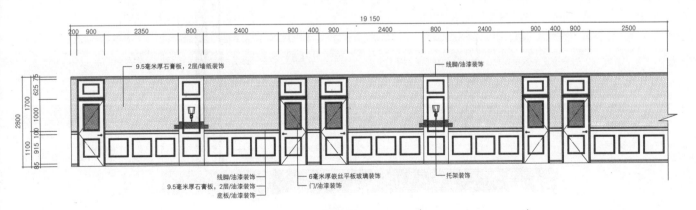

立面图c'

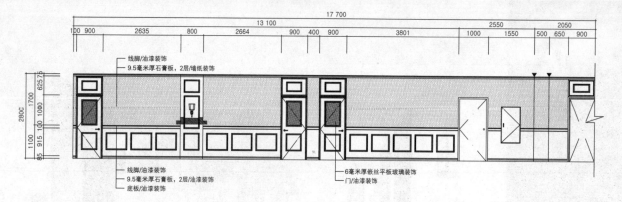

立面图d

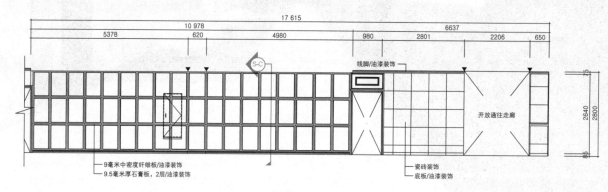

立面图d'

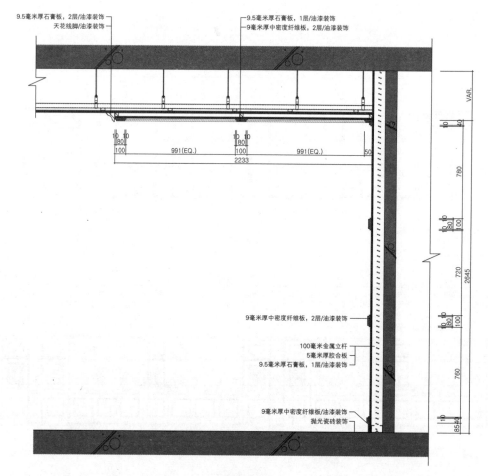

剖面图C

教室

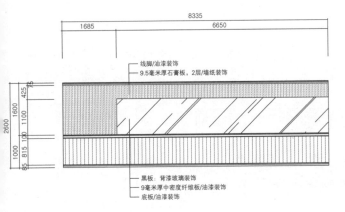

立面图e

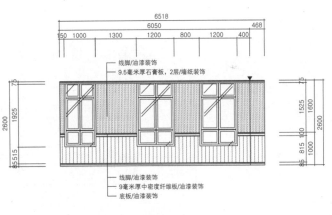

立面图f

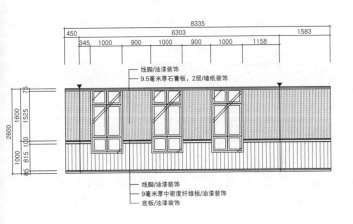

立面图g

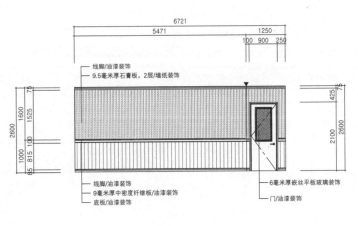

立面图h

研讨室

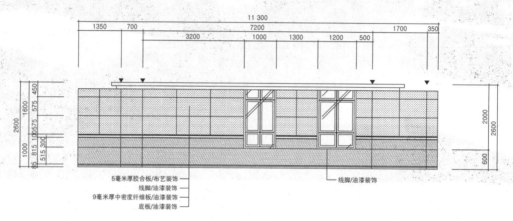

立面图 i

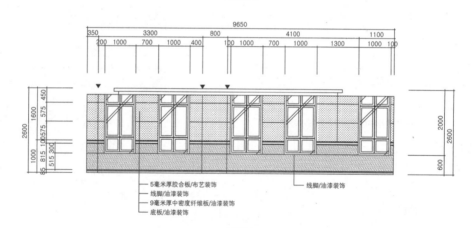

立面图 j

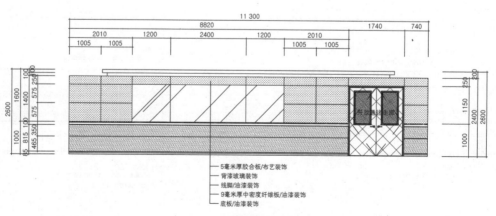

立面图 k

Presby Edu Center, a complex community center with a classical beauty, was newly built. It provides an integral system to learn arts, music, science, dance, and paduk after school.

The overall space reminds of a European school with its classical elements and colors from the logo of a prestige school under the theme of 'Children's Happiness'. The lobby on the underground floor emits coziness through the use of wood, bricks, fireplace, and built-in furniture. The Barrisol lighting compensates for the lack of daylight and also produces a sense of openness and pleasantness. The classical moldings were applied to the furniture and walls on the third floor that houses various facilities for learning. The music room is based on the motive of a piano tune, while the art studio and science room are focused on function. The lecture room and seminar room on the fourth and fifth floors. On the center of the hallway is a library finished with calm blue and a modern pattern, offering a liberal atmosphere for reading and relaxation. Emphasizing the brand identity, the design for Presby Edu is a cozy and comfortable space for children to learn.

Presby 教育中心是新建的一家古典风格的综合社区中心。它提供课余美术、音乐、科学、舞蹈和围棋一体化系统。

以"儿童的幸福"为主题，取自名校校徽的古典元素和颜色令人联想起欧洲的学校。地下楼层大厅通过使用木、砖、壁炉和嵌墙式家具传递出舒适的气息。巴力照明弥补日光的不足，同时营造开放和愉悦的感觉。在设置各种学习设施的第三层，家具和墙壁应用古典线脚。音乐室考虑了钢琴的要素而设计，美术室和科学室同样关注功能性。教室和研讨室布置在第四层和第五层。走廊的中心是令人平和的蓝色和现代化图案装饰的图书室，可在自由的氛围中阅读、放松。Presby 教育中心的设计强调品牌形象，营造儿童学习的舒适空间。

KUT环球教育中心

建筑面积：3828平方米

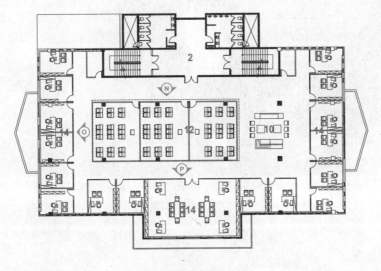

三层平面图

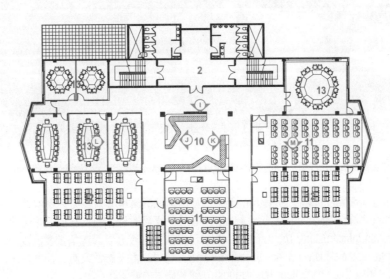

二层平面图

1. 入口
2. 大厅
3. 公共休息区
4. 咖啡店
5. 视听室
6. 大会议室
7. 咨询中心
8. 办公室
9. 主任办公室
10. 休息区
11. 实验室
12. 教室
13. 会议室
14. 教师休息区

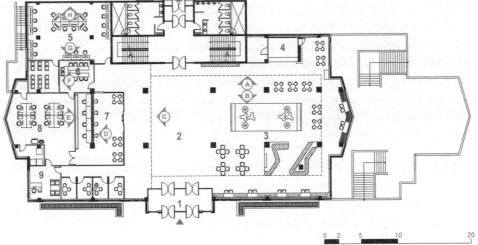

一层平面图

KUT 环球教育中心

The International Education Center a previous library with an underground floor and three above-ground floors, is an educational office space reinforced by language education infrastructure. The facility aims at helping students to improve their language skills which has become a necessary and competitive element today. Thus each floor has various spaces such as global lounge, satellite broadcasting audio-visual room, and lecture rooms for debates and seminars.

The first floor is composed of a main office, information area, cafeteria and an internet bar designed after a navigating ship and the waves of the ocean to allow the users to interchange information, take a rest and have a discussion. The reletively low ceiling height was complemented by breaking down the existing finish to expose the equippments and painting it in dark gray color. Such an approach visually expands the space, and the undulating wood ceiling also adds softness. The cafeteria, which evokes the image of a ship floating on the ocean, and the streamlined booth on the center further enrich the space. The booth penetrated by a white rod has become a symbolic objet of the space.

The second floor, which houses most of the lecture rooms, is the most crowded place in the facility. Thus the designer introduced the color of orange, to create a cheerful and enlivened atmosphere. On the contrary, the third floor, occupied by professors' offices, offers a comfortable and intimate ambiance resulting from the use of brown color and birch plywood for professors and students to have a conversation.

KUT 环球教育中心前身为一家由地下一层和地上三层构成的图书馆，它是经过语言教育基础设施强化的教育办公场所。语言技能已经成为当今必要的竞争要素，该中心旨在帮助学生提高他们的语言能力。每一层都设有公共休息区、卫星广播视听室以及供辩论和研讨的教室等空间。

第一层包括主要办公室、信息区、自助餐厅和一个以航海为主题的网吧，可供使用者交流、休息和讨论。拆除既有装饰，露出设备，并用深灰色上漆，弥补了天花板高度相对较低的弱点。这种做法从视觉上拓展了空间，波浪状的木吊顶增添了一些柔和感。自助餐厅令人联想到海上行船的形象，中间流线型的台位进一步丰富了空间。白杆穿透的台座已经成为场所的象征性符号。

大部分教室设在第二层，是中心最拥挤的地方，设计师采用了橙色，从而营造出一种活泼的气氛。对比之下，第三层包括教授办公室，采用棕色和桦木胶合板，营造出一种舒适、亲密的气氛，便于教授和学生们交谈。

公共休息区的咖啡店

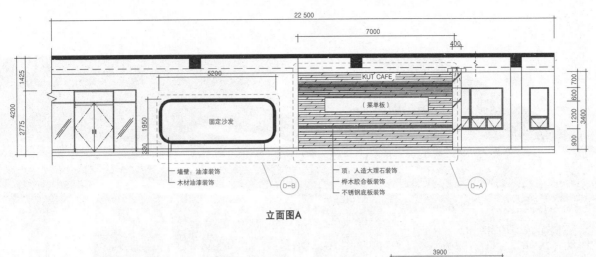

立面图A

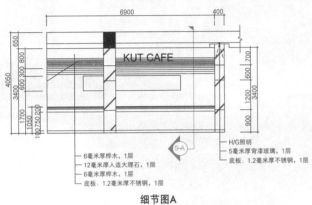

细节图A

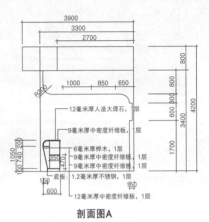

剖面图A

固定沙发

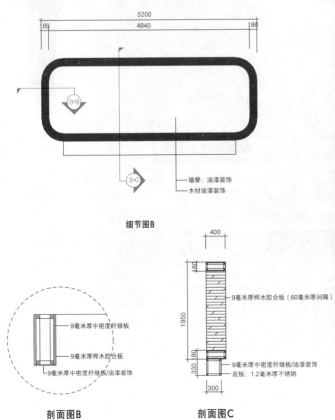

细节图B

剖面图B

剖面图C

KUT 环球教育中心　165

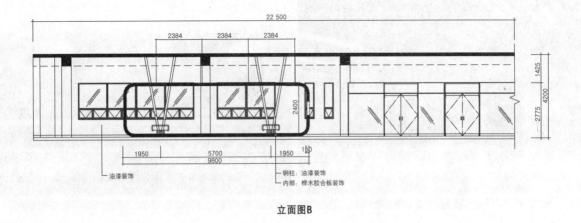

立面图B

圆形隔间

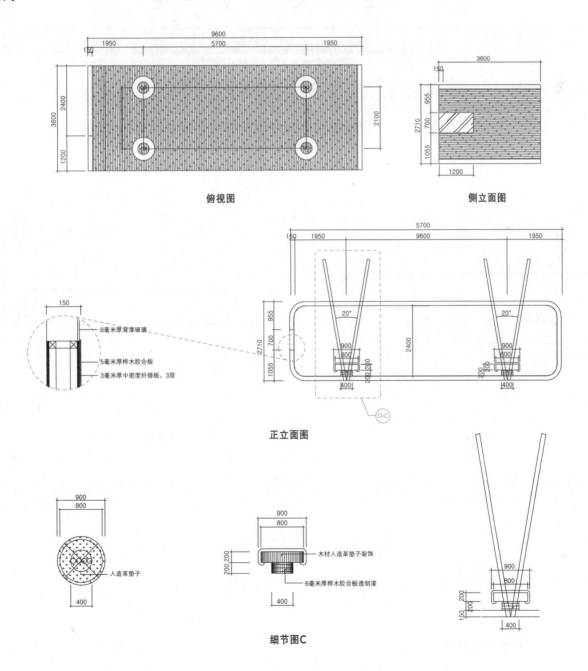

俯视图

侧立面图

正立面图

细节图C

KUT 环球教育中心

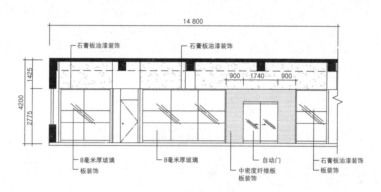

立面图C

■ 咨询中心

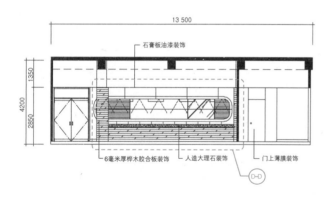

立面图D

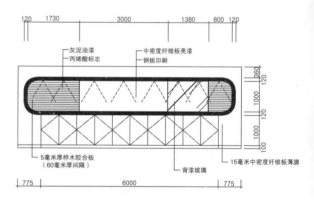

细节图D——前视图

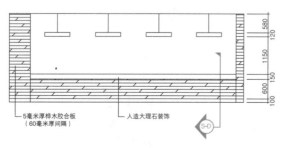

细节图D——后视图

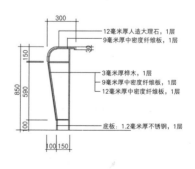

剖面图D

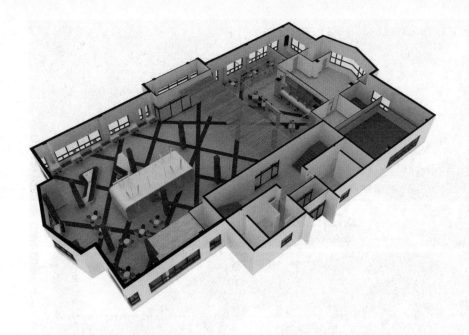

办公室和视听室

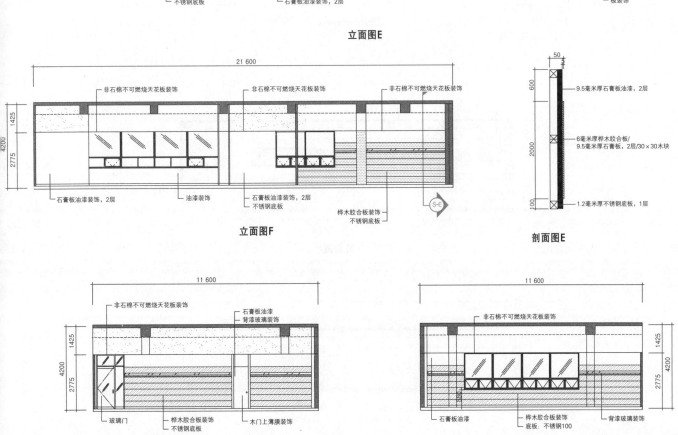

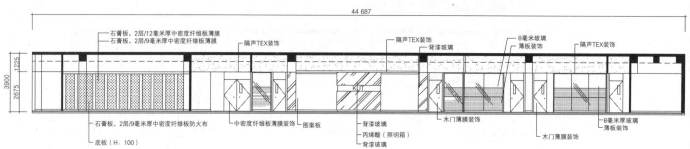

立面图I

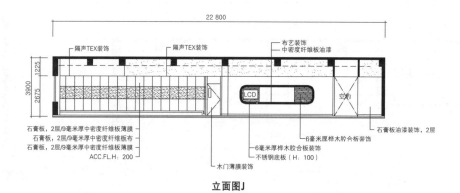

立面图J

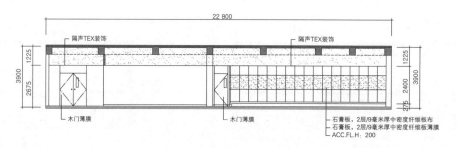

立面图K

第三层休息区和走廊

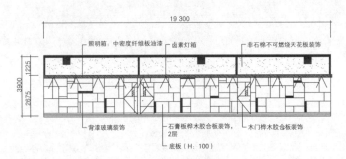

立面图N

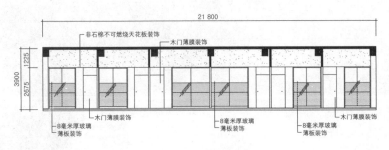

立面图O

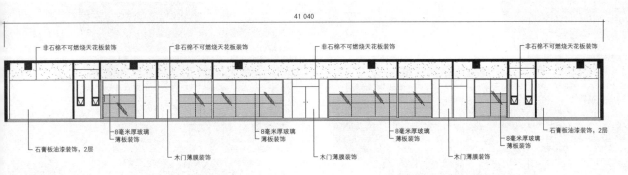

立面图P

KUT 环球教育中心

| 人寿保险 | 建筑面积：3518平方米

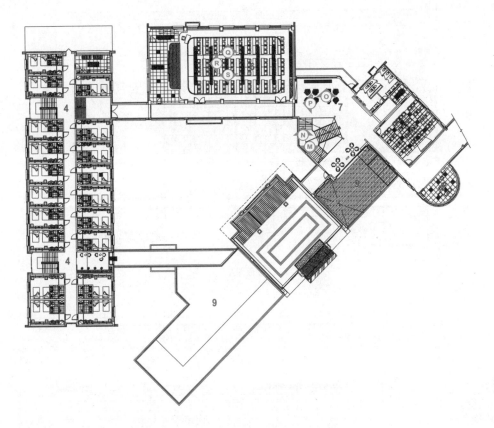

二层平面图

1. 大厅
2. 会议室
3. 网络区
4. 宿舍
5. 院子
6. 教室
7. 休息区
8. 办公室
9. 阳台

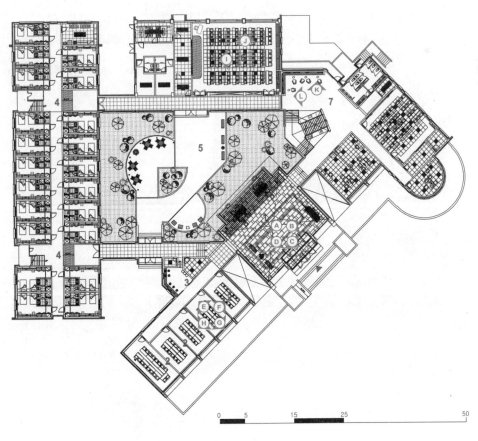

一层平面图

■ 大厅

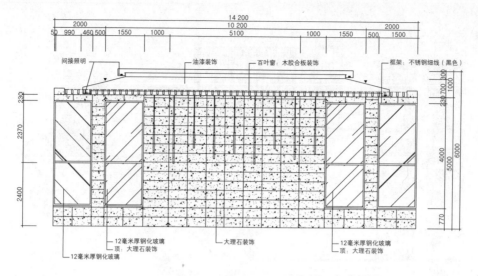

立面图A

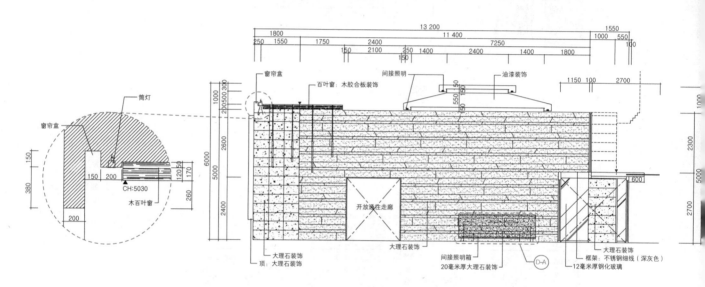

立面图B

■ 咨询台——细节图A

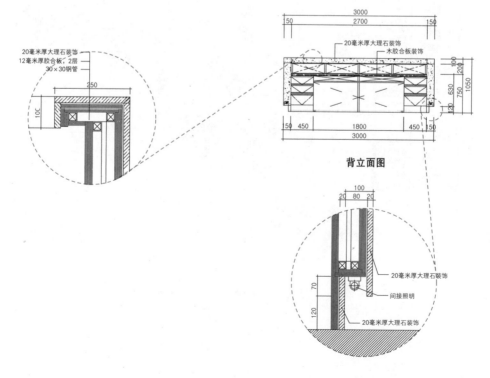

背立面图

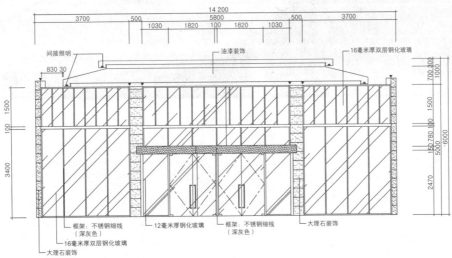

立面图C

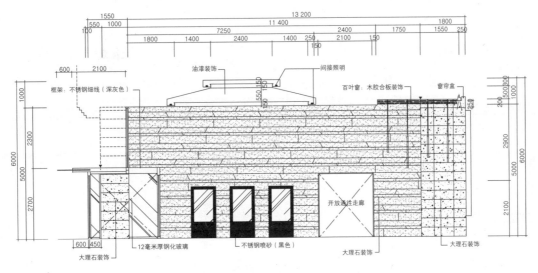

立面图D

会议室

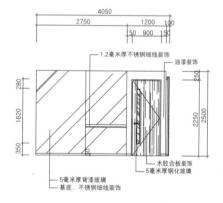

立面图E

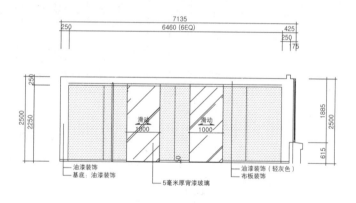

立面图F

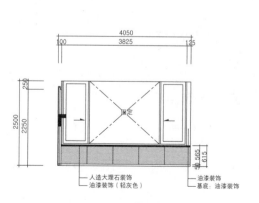

立面图G

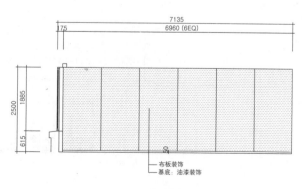

立面图H

第一层楼教室

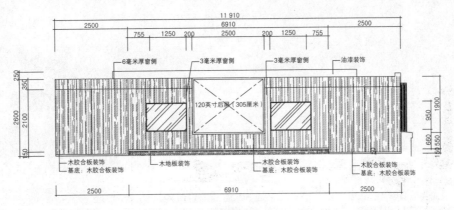

立面图I

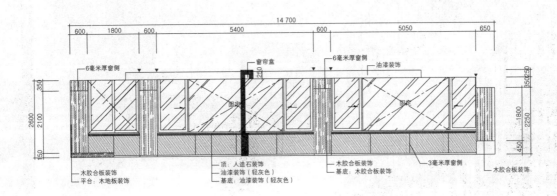

立面图J

一层休息区

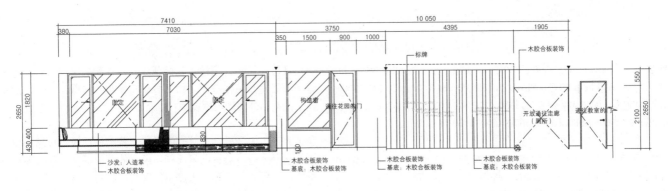

立面图K

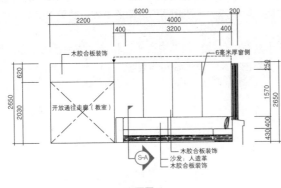

立面图L

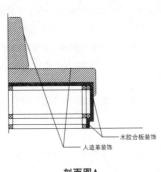

剖面图A

The project involved renovating an old building into a training center for Life Insurance to help its executives and employees to develop their ability. Seeking the harmony with the landscape, we brought changes only to the interior and maintained the bricks of the exterior skin.

Inside the entrance, the lobby continues the atmosphere of the outside to lend a modern and natural feeling. Marble finish on the floor and walls emphasizes the texture, while the existing anterior window ensures the view from the lobby toward the courtyard. The corridor beyond the lobby separates the circulation that leads to guest rooms and education facilities. The wood veneer walls of the corridor is accentuated by typography related to Life Insurance to strengthen the identity of the firm. The lecture room has a round ceiling lined with a sound absorbing material. Each guest room has under-floor heating system, a separate bathroom, a bed and a desk, with the green point wallpaper brightening the small area. Facing the staircase, around the entrance to the second floor is a comfortable, informal lounge where trainees can have a chat or read books.

Respecting the existing environment, the training center effectively represents the company philosophy through the harmony of various elements including the finish material, color, and typography.

该项目旨在将一栋旧建筑物翻新成人寿保险的培训中心，帮助其高管和员工提升工作技能。为了寻求与景观的协调，我们仅对室内进行了改造，外观砖石保持原样。

走进大门，大厅延续了外部的风格，具有一种现代自然的感觉。地面和墙壁大理石装饰凸显质感，既有的前窗提供大厅对外面庭院的视野。大厅后的走廊将人流引向客房和教育设施。走廊的木胶合板墙壁因人寿保险的字样更加显眼，强调了公司的身份。教室采用圆形吊顶，内置隔声材料。所有客房设有地板采暖系统、独立的卫生间、床和书桌，绿色斑点壁纸提亮这个有限的区域。楼梯通向第二层的入口处为一片非正式的舒适休息区，学员可以在此聊天或读书。

关于既有的环境，培训中心通过和谐地使用各种元素，包括装饰材料、颜色和文字传达公司的理念。

楼梯

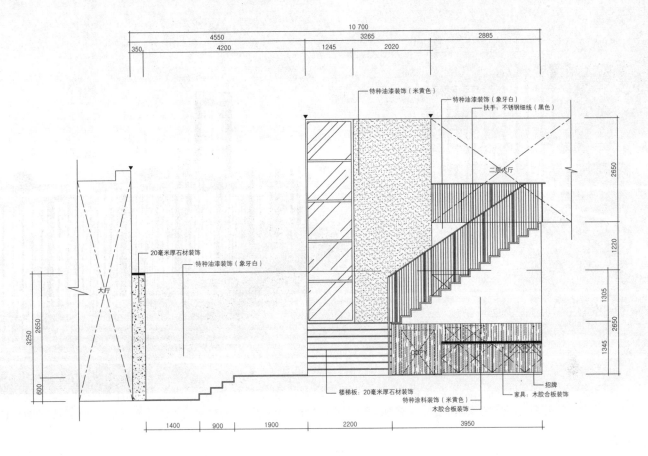

立面图M

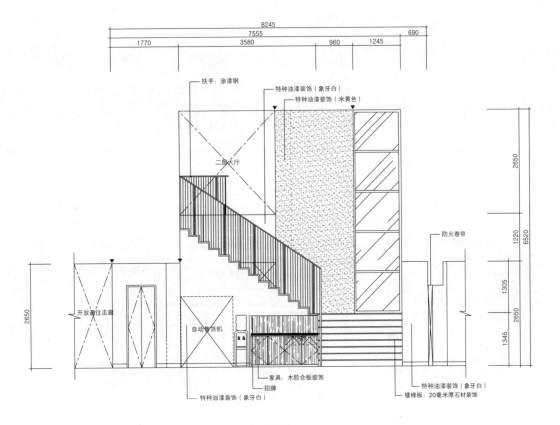

立面图N

■ 二层休息区

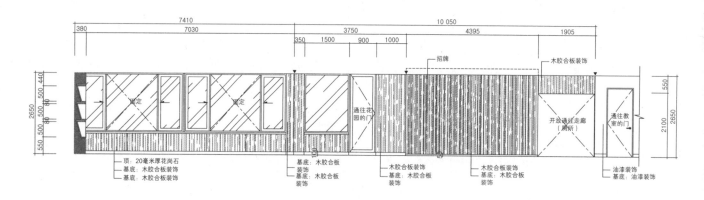

立面图O

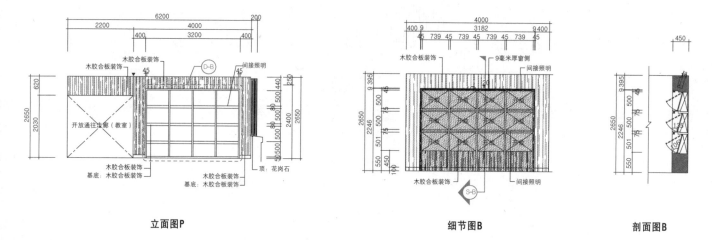

立面图P　　　　　　　　　　细节图B　　　　剖面图B

教室

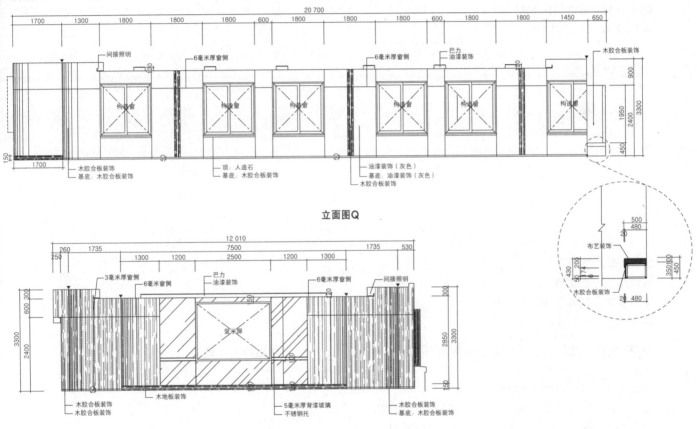

立面图Q

立面图R

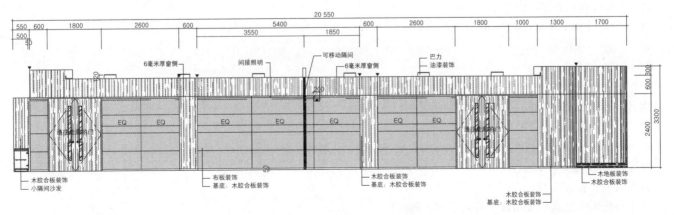

立面图S

图书在版编目（CIP）数据

室内细部图集. 3，办公场所与教育机构 / 凤凰空间编. -- 南昌：江西科学技术出版社，2017.9
ISBN 978-7-5390-5609-8

Ⅰ. ①室… Ⅱ. ①凤… Ⅲ. ①办公室－室内装饰设计－细部设计－图集②教育建筑－室内装饰设计－细部设计－图集 Ⅳ. ①TU238-64

中国版本图书馆CIP数据核字(2017)第115571号

国际互联网(Internet)地址：
http://www.jxkjcbs.com
图书代码：B17041-101
选题序号：KX2017087

责任编辑	魏栋伟
特约编辑	李文恒
项目策划	凤凰空间／单　爽
售后热线	022-87893668

室内细部图集3　办公场所与教育机构　　　　　凤凰空间　编

出版发行	江西科学技术出版社
社　　址	南昌市蓼洲街2号附1号　邮编：330009
	电话：(0791)86623491　86639342(传真)
印　　刷	北京博海升彩色印刷有限公司
经　　销	各地新华书店
开　　本	889 mm×1194 mm　1/16
字　　数	92千
印　　张	11.5
版　　次	2017年9月第1版　2023年3月第2次印刷
书　　号	ISBN 978-7-5390-5609-8
定　　价	198.00元

赣版权登字-03-2017-190
版权所有，侵权必究
（赣科版图书凡属印装错误，可向承印厂调换）